Praise for **T H E S**

"A thought-provoking exploration o̶ ̶ ̶ ̶ ̶ ̶ ̶ ̶
tural consequences, rendering high̶ ̶ ̶ ̶ ̶ ̶ ̶ ̶ ̶ ̶ ̶ ̶ intelligible
to the general reader."   —2011 Pulitzer Prize Committee

"In his new book, *The Shallows*, Nicholas Carr has written a *Silent Spring* for the literary mind."   —Michael Agger, *Slate*

"We are living through something of a backlash against the frenzy of attention dispersion, a backlash for which Carr's book will become canonical."   —Todd Gitlin, *New Republic*

"Carr is a great writer. . . . This is a must-read for any desk jockey concerned about the Web's deleterious effects on the mind. Grade: A."
                                                                        —*Newsweek*

"Editors' choice."   —*New York Times Book Review*

"Carr [is] a Paul Revere for our Net age."   —*USA Today*

"Absorbing [and] disturbing. We all joke about how the Internet is turning us, and especially our kids, into fast-twitch airheads incapable of profound cogitation. It's no joke, Mr. Carr insists, and he has me persuaded."   —John Horgan, *Wall Street Journal*

"I have not only given this book to numerous friends, I actually changed my life in response to it."
                                                        —Jonathan Safran Foer, author of
                                       Extremely Loud and Incredibly Close and *Eating Animals*

"Carr provides a deep, enlightening examination of how the Internet influences the brain and its neural pathways. . . . His fantastic investigation of the effect of the Internet on our neurological selves concludes with a very humanistic petition for balancing our human and computer interactions. . . . Highly recommended."

—*Library Journal*, starred review

"Carr's fresh, lucid, and engaging assessment of our infatuation with the Web is provocative and revelatory."   —*Booklist*

"Mild-mannered, never polemical, with nothing of the Luddite about him, Carr makes his points with a lot of apt citations and wide-ranging erudition."   —Christopher Caldwell, *Financial Times*

"Carr wants us to think deeply about the effects of this new technology on our cultures, our brains, our social lives and our ways of thinking about knowledge. With masterful ease and winning style, he lays out ideas that will encourage readers to do just that. . . . *The Shallows* is a book everyone should read."

—Anna Lena Phillips, *American Scientist*

"Carr's scope in this unceasingly interesting book is wider than just the synapse and the transistor."

—Sam Leith, *Sunday Times* (London)

"Required reading for anyone who wants a cogent, comprehensive, and thoroughly researched statement of the techno-fears that, in however inchoate a way, many of us have harbored for going on a few decades now."   —Daniel Menaker, *Barnes & Noble Review*

"Nicholas Carr has written a deep book about shallow thinking."

—Daniel J. Flynn, *American Spectator*

"If you care about your own ability to think and read deeply, please treat yourself to Carr's book."   —Carol Keeley, *Ploughshares*

"Nicholas Carr's *The Shallows* is a deeply thoughtful, surprising exploration of our 'frenzied' psyches in the age of the Internet. Whether you do it in pixels or pages, read this book."

—Tom Vanderbilt, author of *Traffic*

"Witty, ambitious, and immensely readable, *The Shallows* actually manages to describe the weird, new, artificial world in which we now live." —Dana Gioia, poet and former chairman of the National Endowment for the Arts

"To his great credit, Carr is as even-handed as possible. He consistently emphasizes the fact that screen technologies are neither evil nor miraculous in their effects on the human mind: rather, for every talent lost or diminished, another will be gained or enhanced. What is certain, however, is that our minds will change. . . . *The Shallows* is a worthy illustration that books do, indeed, enable deep reflection."

—Susan Greenfield, *Literary Review*

"Measured but alarming. . . . Carr brilliantly brings together numerous studies and experiments to support this astounding argument: 'With the exception of alphabets and number systems, the Net may well be the single most powerful mind-altering technology that has ever come into general use.'"

—Will Buchanan, *Christian Science Monitor*

"Cogent, urgent and well worth reading." —*Kirkus Reviews*

"*The Shallows* certainly isn't the first examination of this subject, but it's more lucid, concise and pertinent than similar works. . . . An essential, accessible dispatch about how we think now."

—Laura Miller, *Salon*

"The picture of our intellectual future, rendered thoroughly, convincingly, and often beautifully in Carr's text, is bleak enough to give any serious mind some serious pause."—Patrick Tucker, *The Futurist*

"If you retain any residual aspirations for literary repartee, prefer the smell of a book to a mouse and, most important, enjoy the quiet meanderings within your own mind that can be triggered by a good bit of prose, you are the person to whom Nicholas Carr has addressed his riveting new book, *The Shallows*."

—Robert Burton, *San Francisco Chronicle*

"The author of 'Is Google Making Us Stupid?' returns to his thesis at book-length—but can our web-truncated attention spans handle so much prose? With Carr at the wheel, the answer is a resounding 'yes.' . . . *The Shallows* is a guide for understanding—and even regaining control of—your brain on the internet." —*Seed*

"Carr is an excellent writer. One of those nonfiction writers in the league with people like Malcolm Gladwell and Dan Ariely who can teach and entertain at the same time."—Jim Randel, *Huffington Post*

"Outstanding . . . a shrewd, compelling overview of how an ever-changing, always growing technology has changed us." —*BookPage*

"Persuasive. . . . [Carr] cites enough academic research in *The Shallows* to give anyone pause about society's full embrace of the Internet as an unadulterated force for progress."—Peter Burrows, *BusinessWeek*

"Another reason for book lovers not to throw in the towel quite yet is *The Shallows*, by Nicholas Carr, a quietly eloquent retort to those who claim that digital culture is harmless—who claim, in fact, that we're getting smarter by the minute just because we can plug in a computer and allow ourselves to get lost in the funhouse of endless hyperlinks." —Julia Keller, *Chicago Tribune*

THE SHALLOWS

THE SHALLOWS

What the Internet Is Doing
to Our Brains

NICHOLAS CARR

W. W. NORTON & COMPANY
Independent Publishers Since 1923

For information about permission to reproduce selections from this book,
write to Permissions, W. W. Norton & Company, Inc.,
500 Fifth Avenue, New York, NY 10110

For information about special discounts for bulk purchases, please contact
W. W. Norton Special Sales at specialsales@wwnorton.com or 800-233-4830

Manufacturing by LSC Communications, Harrisonburg
Book design by Chris Welch
Production manager: Anna Oler

The Library of Congress has cataloged the hardcover edition as follows:
Carr, Nicholas G., 1959–
The shallows : what the Internet is doing to our brains / Nicholas Carr. — 1st ed.
p. cm.
Includes bibliographical references and index.
ISBN 978-0-393-07222-8 (hardcover)
1. Neuropsychology. 2. Internet—Physiological effect.
3. Internet—Psychological aspects.
I. Title.
QP360.C3667 2010
612.80285—dc22

2010007639

ISBN 978-0-393-35782-0 pbk.

W. W. Norton & Company, Inc.
500 Fifth Avenue, New York, N.Y. 10110
www.wwnorton.com

W. W. Norton & Company Ltd.
15 Carlisle Street, London W1D 3BS

1 2 3 4 5 6 7 8 9 0

to my mother

and in memory of my father

Contents

Introduction to the Second Edition ix

Prologue
THE WATCHDOG AND THE THIEF 1

One
HAL AND ME 5

Two
THE VITAL PATHS 17

a digression
on what the brain thinks about when it thinks
about itself 36

Three
TOOLS OF THE MIND 39

Four
THE DEEPENING PAGE 58

a digression
on lee de forest and his amazing audion   78

Five
A MEDIUM OF THE MOST GENERAL NATURE   81

Six
THE VERY IMAGE OF A BOOK   99

Seven
THE JUGGLER'S BRAIN   115

a digression
on the buoyancy of IQ scores   144

Eight
THE CHURCH OF GOOGLE   149

Nine
SEARCH, MEMORY   177

a digression
on the writing of this book   198

Ten
A THING LIKE ME   201

Epilogue
HUMAN ELEMENTS   223

Afterword to the Second Edition
THE MOST INTERESTING THING IN THE WORLD   225

Notes   239

Further Reading   271

Acknowledgments   275

Index   277

Introduction To the Second Edition

Welcome to *The Shallows*. When I wrote this book ten years ago, the prevailing view of the Internet was sunny, often ecstatically so. We reveled in the seemingly infinite bounties of the online world. We admired the wizards of Silicon Valley and trusted them to act in our best interest. We took it on faith that computer hardware and software would make our lives better, our minds sharper. In a 2010 Pew Research survey of some 400 prominent thinkers, more than eighty percent agreed that "by 2020, people's use of the Internet [will have] enhanced human intelligence; as people are allowed unprecedented access to more information, they become smarter and make better choices."[1]

The year 2020 has arrived. We're not smarter. We're not making better choices.

The Shallows explains why we were mistaken about the Net. When it comes to the quality of our thoughts and judgments, the amount of information a communication medium supplies is less important than the way the medium presents the information and the way, in turn, our minds take it in. The brain's capacity is not unlimited. The passageway from perception to understanding is narrow. It takes patience and concentration to evaluate new information—to gauge its accuracy, to weigh its relevance and worth, to put it into context—and the Internet, by design, subverts patience and concentration. When the brain is overloaded by stimuli, as it usually is when we're peering into a network-connected computer screen, attention splinters, thinking becomes superficial, and memory suffers. We become less reflective and more impulsive. Far from enhancing human intelligence, I argue, the Internet degrades it.

Much has changed in the decade since *The Shallows* came out. Smartphones have become our constant companions. Social media has insinuated itself into everything we do. The dark things that can happen when everyone's connected have happened. Our faith in Silicon Valley has been broken, yet the big Internet companies wield more power than ever. This tenth-anniversary edition of *The Shallows* takes stock of the changes. It includes an extensive new afterword in which I examine the cognitive and cultural consequences of the rise of smartphones and social media, drawing on the large body of new research that has appeared since 2010. I have left the original text of the book largely unchanged. I'm biased, but I think *The Shallows* has aged well. To my eyes, it's more relevant today than it was ten years ago. I hope you find it worthy of your attention.

—NICHOLAS CARR, MASSACHUSETTS, 2020

And in the midst of this wide quietness

A rosy sanctuary will I dress

With the wreath'd trellis of a working brain . . .

— JOHN KEATS, "Ode to Psyche"

THE SHALLOWS

Prologue

THE WATCHDOG AND THE THIEF

I n 1964, just as the Beatles were launching their invasion of America's airwaves, Marshall McLuhan published *Understanding Media: The Extensions of Man* and transformed himself from an obscure academic into a star. Oracular, gnomic, and mindbending, the book was a perfect product of the sixties, that nowdistant decade of acid trips and moon shots, inner and outer voyaging. *Understanding Media* was at heart a prophecy, and what it prophesied was the dissolution of the linear mind. McLuhan declared that the "electric media" of the twentieth century—telephone, radio, movies, television—were breaking the tyranny of text over our thoughts and senses. Our isolated, fragmented selves, locked for centuries in the private reading of printed pages, were becoming whole again, merging into the global equivalent of a tribal village. We were approaching "the technological simulation of consciousness, when the creative process of knowing will be collectively and corporately extended to the whole of human society."[1]

Even at the crest of its fame, *Understanding Media* was a book more talked about than read. Today it has become a cultural relic, consigned to media studies courses in universities. But McLuhan, as much a showman as a scholar, was a master at turning phrases, and one of them, sprung from the pages of the book, lives on as a popular saying: "The medium is the message." What's been forgotten in our repetition of this enigmatic aphorism is that McLuhan was not just acknowledging, and celebrating, the transformative power of new communication technologies. He was also sounding a warning about the threat the power poses—and the risk of being oblivious to that threat. "The electric technology is within the gates," he wrote, "and we are numb, deaf, blind and mute about its encounter with the Gutenberg technology, on and through which the American way of life was formed."[2]

McLuhan understood that whenever a new medium comes along, people naturally get caught up in the information—the "content"—it carries. They care about the news in the newspaper, the music on the radio, the shows on the TV, the words spoken by the person on the far end of the phone line. The technology of the medium, however astonishing it may be, disappears behind whatever flows through it—facts, entertainment, instruction, conversation. When people start debating (as they always do) whether the medium's effects are good or bad, it's the content they wrestle over. Enthusiasts celebrate it; skeptics decry it. The terms of the argument have been pretty much the same for every new informational medium, going back at least to the books that came off Gutenberg's press. Enthusiasts, with good reason, praise the torrent of new content that the technology uncorks, seeing it as signaling a "democratization" of culture. Skeptics, with equally good reason, condemn the crassness of the content, viewing it as signaling a "dumbing down" of culture. One side's abundant Eden is the other's vast wasteland.

The Internet is the latest medium to spur this debate. The clash between Net enthusiasts and Net skeptics, carried out over the last two decades through dozens of books and articles and thousands of blog posts, video clips, and podcasts, has become as polarized as ever, with the former heralding a new golden age of access and participation and the latter bemoaning a new dark age of mediocrity and narcissism. The debate has been important—content does matter—but because it hinges on personal ideology and taste, it has gone down a cul-de-sac. The views have become extreme, the attacks personal. "Luddite!" sneers the enthusiast. "Philistine!" scoffs the skeptic. "Cassandra!" "Pollyanna!"

What both enthusiast and skeptic miss is what McLuhan saw: that in the long run a medium's content matters less than the medium itself in influencing how we think and act. As our window onto the world, and onto ourselves, a popular medium molds what we see and how we see it—and eventually, if we use it enough, it changes who we are, as individuals and as a society. "The effects of technology do not occur at the level of opinions or concepts," wrote McLuhan. Rather, they alter "patterns of perception steadily and without any resistance."[3] The showman exaggerates to make his point, but the point stands. Media work their magic, or their mischief, on the nervous system itself.

Our focus on a medium's content can blind us to these deep effects. We're too busy being dazzled or disturbed by the programming to notice what's going on inside our heads. In the end, we come to pretend that the technology itself doesn't matter. It's how we use it that matters, we tell ourselves. The implication, comforting in its hubris, is that we're in control. The technology is just a tool, inert until we pick it up and inert again once we set it aside.

McLuhan quoted a self-serving pronouncement by David Sarnoff, the media mogul who pioneered radio at RCA and television at NBC. In a speech at the University of Notre Dame in 1955, Sarnoff dismissed criticism of the mass media on which he had built his empire and his fortune. He turned the blame for any ill effects away from the technologies and onto the listeners and viewers: "We are too prone to make technological instruments the scapegoats for the sins of those who wield them. The products of modern science are not in themselves good or bad; it is the way they are used that determines their value." McLuhan scoffed at the idea, chiding Sarnoff for speaking with "the voice of the current somnambulism."[4]

Every new medium, McLuhan understood, changes us. "Our conventional response to all media, namely that it is how they are used that counts, is the numb stance of the technological idiot," he wrote. The content of a medium is just "the juicy piece of meat carried by the burglar to distract the watchdog of the mind."[5]

Not even McLuhan could have foreseen the feast that the Internet has laid before us: one course after another, each juicier than the last, with hardly a moment to catch our breath between bites. As networked computers have shrunk to the size of iPhones and Androids, the feast has become a movable one, available anytime, anywhere. It's in our home, our office, our car, our classroom, our purse, our pocket. Even people who are wary of the Net's ever-expanding influence rarely allow their concerns to get in the way of their use and enjoyment of the technology. The movie critic David Thomson once observed that "doubts can be rendered feeble in the face of the certainty of the medium."[6] He was talking about the cinema and how it projects its sensations and sensibilities not only onto the movie screen but onto us, the engrossed and compliant audience. His comment applies with even greater force to the Net. The computer screen bulldozes our doubts with its bounties and conveniences. It is so much our servant that it would seem churlish to notice that it is also our master.

One

HAL AND ME

"Dave, stop. Stop, will you? Stop, Dave. Will you stop?" So the supercomputer HAL pleads with the implacable astronaut Dave Bowman in a famous and weirdly poignant scene toward the end of Stanley Kubrick's *2001: A Space Odyssey.* Bowman, having nearly been sent to a deep-space death by the malfunctioning machine, is calmly, coldly disconnecting the memory circuits that control its artificial brain. "Dave, my mind is going," HAL says, forlornly. "I can feel it. I can feel it."

I can feel it too. Over the last few years I've had an uncomfortable sense that someone, or something, has been tinkering with my brain, remapping the neural circuitry, reprogramming the memory. My mind isn't going—so far as I can tell—but it's changing. I'm not thinking the way I used to think. I feel it most strongly when I'm reading. I used to find it easy to immerse myself in a book or a lengthy article. My mind would get caught up in the twists of the narrative or the turns of the argument, and I'd spend hours strolling through long stretches of prose. That's rarely the case anymore. Now my concentration starts to drift after a page or two. I get fidgety, lose the thread, begin looking for something else to do. I feel like I'm

always dragging my wayward brain back to the text. The deep reading that used to come naturally has become a struggle.

I think I know what's going on. For well over a decade now, I've
been spending a lot of time online, searching and surfing and
sometimes adding to the great databases of the Internet. The Web's
been a godsend to me as a writer. Research that once required days
in the stacks or periodical rooms of libraries can now be done in
minutes. A few Google searches, some quick clicks on hyperlinks,
and I've got the telltale fact or the pithy quote I was after. I couldn't
begin to tally the hours or the gallons of gasoline the Net has saved
me. I do most of my banking and a lot of my shopping online. I use
my browser to pay my bills, schedule my appointments, book flights
and hotel rooms, renew my driver's license, send invitations and
greeting cards. Even when I'm not working, I'm as likely as not to be
foraging in the Web's data thickets—reading and writing e-mails,
scanning headlines and blog posts, following Facebook updates,
watching video streams, downloading music, or just tripping lightly
from link to link to link.

The Net has become my all-purpose medium, the conduit for
most of the information that flows through my eyes and ears and
into my mind. The advantages of having immediate access to such
an incredibly rich and easily searched store of data are many, and
they've been widely described and duly applauded. "Google," says
Heather Pringle, a writer with *Archaeology* magazine, "is an astonishing boon to humanity, gathering up and concentrating information and ideas that were once scattered so broadly around the world
that hardly anyone could profit from them."[1] Observes *Wired*'s Clive
Thompson, "The perfect recall of silicon memory can be an enormous boon to thinking."[2]

The boons are real. But they come at a price. As McLuhan suggested, media aren't just channels of information. They supply the
stuff of thought, but they also shape the process of thought. And
what the Net seems to be doing is chipping away my capacity for concentration and contemplation. Whether I'm online or not, my mind

now expects to take in information the way the Net distributes it: in a swiftly moving stream of particles. Once I was a scuba diver in the sea of words. Now I zip along the surface like a guy on a Jet Ski.

Maybe I'm an aberration, an outlier. But it doesn't seem that way. When I mention my troubles with reading to friends, many say they're suffering from similar afflictions. The more they use the Web, the more they have to fight to stay focused on long pieces of writing. Some worry they're becoming chronic scatterbrains. Several of the bloggers I follow have also mentioned the phenomenon. Scott Karp, who used to work for a magazine and now writes a blog about online media, confesses that he has stopped reading books altogether. "I was a lit major in college, and used to be [a] voracious book reader," he writes. "What happened?" He speculates on the answer: "What if I do all my reading on the web not so much because the way I read has changed, i.e. I'm just seeking convenience, but because the way I THINK has changed?"[3]

Bruce Friedman, who blogs about the use of computers in medicine, has also described how the Internet is altering his mental habits. "I now have almost totally lost the ability to read and absorb a longish article on the web or in print," he says.[4] A pathologist on the faculty of the University of Michigan Medical School, Friedman elaborated on his comment in a telephone conversation with me. His thinking, he said, has taken on a "staccato" quality, reflecting the way he quickly scans short passages of text from many sources online. "I can't read *War and Peace* anymore," he admitted. "I've lost the ability to do that. Even a blog post of more than three or four paragraphs is too much to absorb. I skim it."

Philip Davis, a doctoral student in communication at Cornell who contributes to the Society for Scholarly Publishing's blog, recalls a time back in the 1990s when he showed a friend how to use a Web browser. He says he was "astonished" and "even irritated" when the woman paused to read the text on the sites she stumbled upon. "You're not supposed to read web pages, just click on the hypertexted words!" he scolded her. Now, Davis writes, "I read a lot—or at least

I should be reading a lot—only I don't. I skim. I scroll. I have very little patience for long, drawn-out, nuanced arguments, even though I accuse others of painting the world too simply."[5]

Karp, Friedman, and Davis—all well-educated men with a keenness for writing—seem fairly sanguine about the decay of their faculties for reading and concentrating. All things considered, they say, the benefits they get from using the Net—quick access to loads of information, potent searching and filtering tools, an easy way to share their opinions with a small but interested audience—make up for the loss of their ability to sit still and turn the pages of a book or a magazine. Friedman told me, in an e-mail, that he's "never been more creative" than he has been recently, and he attributes that "to my blog and the ability to review/scan 'tons' of information on the web." Karp has come to believe that reading lots of short, linked snippets online is a more efficient way to expand his mind than reading "250-page books," though, he says, "we can't yet recognize the superiority of this networked thinking process because we're measuring it against our old linear thought process."[6] Muses Davis, "The Internet may have made me a less patient reader, but I think that in many ways, it has made me smarter. More connections to documents, artifacts, and people means more external influences on my thinking and thus on my writing."[7] All three know they've sacrificed something important, but they wouldn't go back to the way things used to be.

For some people, the very idea of reading a book has come to seem old-fashioned, maybe even a little silly—like sewing your own shirts or butchering your own meat. "I don't read books," says Joe O'Shea, a former president of the student body at Florida State University and a 2008 recipient of a Rhodes Scholarship. "I go to Google, and I can absorb relevant information quickly." O'Shea, a philosophy major, doesn't see any reason to plow through chapters of text when it takes but a minute or two to cherry-pick the pertinent passages using Google Book Search. "Sitting down and going through a book from cover to cover doesn't make sense," he says. "It's not a good use of my time, as I can get all the information I need faster through the Web." As soon as you learn to be "a skilled hunter" online, he argues, books become superfluous.[8]

O'Shea seems more the rule than the exception. In 2008, a research and consulting outfit called nGenera released a study of the effects of Internet use on the young. The company interviewed some six thousand members of what it calls "Generation Net"—kids who have grown up using the Web. "Digital immersion," wrote the lead researcher, "has even affected the way they absorb information. They don't necessarily read a page from left to right and from top to bottom. They might instead skip around, scanning for pertinent information of interest."[9] In a talk at a recent Phi Beta Kappa meeting, Duke University professor Katherine Hayles confessed, "I can't get my students to read whole books anymore."[10] Hayles teaches English; the students she's talking about are students of literature.

People use the Internet in all sorts of ways. Some are eager, even compulsive adopters of the latest technologies. They keep accounts with a dozen or more online services and subscribe to scores of information feeds. They post and they comment, they text and they tweet. Others don't much care about being on the cutting edge but nevertheless find themselves online most of the time, tapping away at their desktop, their laptop, or their phone. The Net has become essential to their work, school, or social lives, and often to all three. Still others log on only a few times a day—to check their e-mail, follow a story in the news, research a topic of interest, or do some shopping. And there are, of course, many people who don't use the Internet at all, either because they can't afford to or because they don't want to. What's clear, though, is that for society as a whole the Net has become, in just the twenty years since the software programmer Tim Berners-Lee wrote the code for the World Wide Web, the communication and information medium of choice. The scope of its use is unprecedented, even by the standards of the mass media of the twentieth century. The scope of its influence is equally broad. By choice or necessity, we've embraced the Net's uniquely rapid-fire mode of collecting and dispensing information.

We seem to have arrived, as McLuhan said we would, at an important juncture in our intellectual and cultural history, a moment of transition between two very different modes of thinking. What we're trading away in return for the riches of the Net—and only a curmudgeon would refuse to see the riches—is what Karp calls "our old linear thought process." Calm, focused, undistracted, the linear mind is being pushed aside by a new kind of mind that wants and needs to take in and dole out information in short, disjointed, often overlapping bursts—the faster, the better. John Battelle, a onetime magazine editor and journalism professor who now runs an online advertising syndicate, has described the intellectual frisson he experiences when skittering across Web pages: "When I am performing bricolage in real time over the course of hours, I am 'feeling' my brain light up, I [am] 'feeling' like I'm getting smarter."[11] Most of us have experienced similar sensations while online. The feelings are intoxicating—so much so that they can distract us from the Net's deeper cognitive consequences.

For the last five centuries, ever since Gutenberg's printing press made book reading a popular pursuit, the linear, literary mind has been at the center of art, science, and society. As supple as it is subtle, it's been the imaginative mind of the Renaissance, the rational mind of the Enlightenment, the inventive mind of the Industrial Revolution, even the subversive mind of Modernism. It may soon be yesterday's mind.

THE HAL 9000 computer was born, or "made operational," as HAL himself humbly put it, on January 12, 1992, in a mythical computer plant in Urbana, Illinois. I was born almost exactly thirty-three years earlier, in January of 1959, in another midwestern city, Cincinnati, Ohio. My life, like the lives of most Baby Boomers and Generation Xers, has unfolded like a two-act play. It opened with Analogue Youth and then, after a quick but thorough shuffling of the props, it entered Digital Adulthood.

When I summon up images from my early years, they seem at once comforting and alien, like stills from a G-rated David Lynch film. There's the bulky mustard-yellow telephone affixed to the wall of our kitchen, with its rotary dial and long, coiled cord. There's my dad fiddling with the rabbit ears on top of the TV, vainly trying to get rid of the snow obscuring the Reds game. There's the rolled-up, dew-dampened morning newspaper lying in our gravel driveway. There's the hi-fi console in the living room, a few record jackets and dust sleeves (some from my older siblings' Beatles albums) scattered on the carpet around it. And downstairs, in the musty basement family room, there are the books on the bookshelves—lots of books—with their many-colored spines, each bearing a title and the name of a writer.

In 1977, the year *Star Wars* came out and the Apple Computer company was incorporated, I headed to New Hampshire to attend Dartmouth College. I didn't know it when I applied, but Dartmouth had long been a leader in academic computing, playing a pivotal role in making the power of data-processing machines easily available to students and teachers. The college's president, John Kemeny, was a respected computer scientist who in 1972 had written an influential book called *Man and the Computer*. He had also, a decade before that, been one the inventors of BASIC, the first programming language to use common words and everyday syntax. Near the center of the school's grounds, just behind the neo-Georgian Baker Library with its soaring bell tower, squatted the single-story Kiewit Computation Center, a drab, vaguely futuristic concrete building that housed the school's pair of General Electric GE-635 mainframe computers. The mainframes ran the groundbreaking Dartmouth Time-Sharing System, an early type of network that allowed dozens of people to use the computers simultaneously. Time-sharing was the first manifestation of what we today call personal computing. It made possible, as

Kemeny wrote in his book, "a true symbiotic relationship between man and computer."[12]

I was an English major and went to great lengths to avoid math and science classes, but Kiewit occupied a strategic location on campus, midway between my dorm and Fraternity Row, and on weekend evenings I'd often spend an hour or two at a terminal in the public teletype room while waiting for the keg parties to get rolling. Usually, I'd fritter away the time playing one of the goofily primitive multiplayer games that the undergraduate programmers—"sysprogs," they called themselves—had hacked together. But I did manage to teach myself how to use the system's cumbersome word-processing program and even learned a few BASIC commands.

That was just a digital dalliance. For every hour I passed in Kiewit, I must have spent two dozen next door in Baker. I crammed for exams in the library's cavernous reading room, looked up facts in the weighty volumes on the reference shelves, and worked part-time checking books in and out at the circulation desk. Most of my library time, though, went to wandering the long, narrow corridors of the stacks. Despite being surrounded by tens of thousands of books, I don't remember feeling the anxiety that's symptomatic of what we today call "information overload." There was something calming in the reticence of all those books, their willingness to wait years, decades even, for the right reader to come along and pull them from their appointed slots. *Take your time*, the books whispered to me in their dusty voices. *We're not going anywhere.*

It was in 1986, five years after I left Dartmouth, that computers entered my life in earnest. To my wife's dismay, I spent nearly our entire savings, some $2,000, on one of Apple's earliest Macintoshes—a Mac Plus decked out with a single megabyte of RAM, a 20-megabyte hard drive, and a tiny black-and-white screen. I still recall the excitement I felt as I unpacked the little beige machine. I set it on my desk, plugged in the keyboard and mouse, and flipped the power switch. It lit up, sounded a welcoming chime, and smiled at me as it went through the mysterious routines that brought it to life. I was smitten.

The Plus did double duty as both a home and a business computer. Every day, I lugged it into the offices of the management consulting firm where I worked as an editor. I used Microsoft Word to revise proposals, reports, and presentations, and sometimes I'd launch Excel to key in revisions to a consultant's spreadsheet. Every evening, I carted it back home, where I used it to keep track of the family finances, write letters, play games (still goofy, but less primitive), and—most diverting of all—cobble together simple databases using the ingenious HyperCard application that back then came with every Mac. Created by Bill Atkinson, one of Apple's most inventive programmers, HyperCard incorporated a hypertext system that anticipated the look and feel of the World Wide Web. Where on the Web you click links on pages, on HyperCard you clicked buttons on cards—but the idea, and its seductiveness, was the same.

The computer, I began to sense, was more than just a simple tool that did what you told it to do. It was a machine that, in subtle but unmistakable ways, exerted an influence over you. The more I used it, the more it altered the way I worked. At first I had found it impossible to edit anything on-screen. I'd print out a document, mark it up with a pencil, and type the revisions back into the digital version. Then I'd print it out again and take another pass with the pencil. Sometimes I'd go through the cycle a dozen times a day. But at some point—and abruptly—my editing routine changed. I found I could no longer write or revise anything on paper. I felt lost without the Delete key, the scrollbar, the cut and paste functions, the Undo command. I *had* to do all my editing on-screen. In using the word processor, I had become something of a word processor myself.

Bigger changes came after I bought a modem, sometime around 1990. Up to then, the Plus had been a self-contained machine, its functions limited to whatever software I installed on its hard drive. When hooked up to other computers through the modem, it took on a new identity and a new role. It was no longer just a high-tech Swiss Army knife. It was a communications medium, a device for finding, organizing, and sharing information. I tried all the online services— CompuServe, Prodigy, even Apple's short-lived eWorld—but the one I stuck with was America Online. My original AOL subscription limited me to five hours online a week, and I would painstakingly parcel out the precious minutes to exchange e-mails with a small group of friends who also had AOL accounts, to follow the conversations on a few bulletin boards, and to read articles reprinted from newspapers and magazines. I actually grew fond of the sound of my modem connecting through the phone lines to the AOL servers. Listening to the bleeps and clangs was like overhearing a friendly argument between a couple of robots.

By the mid-nineties, I had become trapped, not unhappily, in the "upgrade cycle." I retired the aging Plus in 1994, replacing it with a Macintosh Performa 550 with a color screen, a CD-ROM drive, a 500-megabyte hard drive, and what seemed at the time a miraculously fast 33-megahertz processor. The new computer required updated versions of most of the programs I used, and it let me run all sorts of new applications with the latest multimedia features. By the time I had installed all the new software, my hard drive was full. I had to go out and buy an external drive as a supplement. I added a Zip drive too—and then a CD burner. Within a couple of years, I'd bought another new desktop, with a much larger monitor and a much faster chip, as well as a portable model that I could use while traveling. My employer had, in the meantime, banished Macs in favor of Windows PCs, so I was using two different systems, one at work and one at home.

It was around this same time that I started hearing talk of something called the Internet, a mysterious "network of networks" that promised, according to people in the know, to "change everything." A 1994 article in Wired declared my beloved AOL "suddenly obsolete." A new invention, the "graphical browser," promised a far more exciting digital experience: "By following the links—click, and the linked document appears—you can travel through the online world along paths of whim and intuition."[13] I was intrigued, and then I was hooked. By the end of 1995 I had installed the new Netscape browser on my work computer and was using it to explore the seemingly infinite pages of the World Wide Web. Soon I had an ISP account at home as well—and a much faster modem to go with it. I canceled my AOL service.

You know the rest of the story because it's probably your story too. Ever-faster chips. Ever-quicker modems. DVDs and DVD burners. Gigabyte-sized hard drives. Yahoo and Amazon and eBay. MP3s. Streaming video. Broadband. Napster and Google. BlackBerrys and iPods. Wi-Fi networks. YouTube and Wikipedia. Blogging and microblogging. Smartphones, thumb drives, netbooks. Who could resist? Certainly not I.

When the Web went 2.0 around 2005, I went 2.0 with it. I became a social networker and a content generator. I registered a domain, roughtype.com, and launched a blog. It was exhilarating, at least for the first couple of years. I had been working as a freelance writer since the start of the decade, writing mainly about technology, and I knew that publishing an article or a book was a slow, involved, and often frustrating business. You slaved over a manuscript, sent it off to a publisher, and, assuming it wasn't sent back with a rejection slip, went through rounds of editing, fact checking, and proofreading. The finished product wouldn't appear until weeks or months later. If it was a book, you might have to wait more than a year to see it in print. Blogging junked the traditional publishing apparatus. You'd type something up, code a few links, hit the Publish button, and your work would be out there, immediately, for all the world to see. You'd also get something you rarely got with more formal writing: direct responses from readers, in the form of comments or, if the readers had their own blogs, links. It felt new and liberating.

Reading online felt new and liberating too. Hyperlinks and search engines delivered an endless supply of words to my screen, alongside pictures, sounds, and videos. As publishers tore down their paywalls, the flood of free content turned into a tidal wave. Headlines streamed around the clock through my Yahoo home page and my RSS feed reader. One click on a link led to a dozen or a hundred more. New e-mails popped into my in-box every minute or two. I registered for accounts with MySpace and Facebook, Digg and Twitter. I started letting my newspaper and magazine subscriptions lapse. Who needed them? By the time the print editions arrived, dew-dampened or otherwise, I felt like I'd already seen all the stories.

Sometime in 2007, a serpent of doubt slithered into my info-paradise. I began to notice that the Net was exerting a much stronger and broader influence over me than my old stand-alone PC ever had. It wasn't just that I was spending so much time staring into a computer screen. It wasn't just that so many of my habits and routines were changing as I became more accustomed to and dependent on the sites and services of the Net. The very way my brain worked seemed to be changing. It was then that I began worrying about my inability to pay attention to one thing for more than a couple of minutes. At first I'd figured that the problem was a symptom of middle-age mind rot. But my brain, I realized, wasn't just drifting. It was hungry. It was demanding to be fed the way the Net fed it—and the more it was fed, the hungrier it became. Even when I was away from my computer, I yearned to check e-mail, click links, do some Googling. I wanted to be *connected*. Just as Microsoft Word had turned me into a flesh-and-blood word processor, the Internet, I sensed, was turning me into something like a high-speed data-processing machine, a human HAL.

I missed my old brain.

Two

THE VITAL PATHS

Friedrich Nietzsche was desperate. Sickly as a child, he had never fully recovered from injuries he suffered in his early twenties when he fell from a horse while serving in a mounted artillery unit in the Prussian army. In 1879, his health problems worsening, he'd been forced to resign his post as a professor of philology at the University of Basel. Just thirty-four years old, he began to wander through Europe, seeking relief from his many ailments. He would head south to the shores of the Mediterranean when the weather turned cool in the fall, then north again, to the Swiss Alps or his mother's home near Leipzig, in the spring. Late in 1881, he rented a garret apartment in the Italian port city of Genoa. His vision was failing, and keeping his eyes focused on a page had become exhausting and painful, often bringing on crushing headaches and fits of vomiting. He'd been forced to curtail his writing, and he feared he would soon have to give it up.

At wit's end, he ordered a typewriter—a Danish-made Malling-Hansen Writing Ball—and it was delivered to his lodgings during the first weeks of 1882. Invented a few years earlier by Hans Rasmus Johann Malling-Hansen, the principal of the Royal Institute for the

Deaf-Mute in Copenhagen, the writing ball was an oddly beautiful instrument. It resembled an ornate golden pincushion. Fifty-two keys, for capital and lowercase letters as well as numerals and punctuation marks, protruded from the top of the ball in a concentric arrangement scientifically designed to enable the most efficient typing possible. Directly below the keys lay a curved plate that held a sheet of typing paper. Using an ingenious gearing system, the plate advanced like clockwork with each stroke of a key. Given enough practice, a person could type as many as eight hundred characters a minute with the machine, making it the fastest typewriter that had ever been built.[1]

The writing ball rescued Nietzsche, at least for a time. Once he had learned touch typing, he was able to write with his eyes closed, using only the tips of his fingers. Words could again pass from his mind to the page. He was so taken with Malling-Hansen's creation that he typed up a little ode to it:

> The writing ball is a thing like me: made of iron
> Yet easily twisted on journeys.
> Patience and tact are required in abundance,
> As well as fine fingers, to use us.

In March, a Berlin newspaper reported that Nietzsche "feels better than ever" and, thanks to his typewriter, "has resumed his writing activities."

But the device had a subtler effect on his work. One of Nietzsche's closest friends, the writer and composer Heinrich Köselitz, noticed a change in the style of his writing. Nietzsche's prose had become tighter, more telegraphic. There was a new forcefulness to it, too, as though the machine's power—its "iron"—was, through some mysterious metaphysical mechanism, being transferred into the words it pressed into the page. "Perhaps you will through this instrument even take to a new idiom," Köselitz wrote in a letter, noting that, in his own work, "my 'thoughts' in music and language often depend on the quality of pen and paper."

"You are right," Nietzsche replied. "Our writing equipment takes part in the forming of our thoughts."[2]

WHILE NIETZSCHE WAS learning to type on his writing ball in Genoa, five hundred miles to the northeast a young medical student named Sigmund Freud was working as a neurophysiology researcher in a Vienna laboratory. His specialty was dissecting the nervous systems of fish and crustaceans. Through his experiments, he came to surmise that the brain, like other bodily organs, is made up of many separate cells. He later extended his theory to suggest that the gaps between the cells—the "contact barriers," as he termed them—play an essential role in governing the functions of the mind, shaping our memories and our thoughts. At the time, Freud's conclusions lay outside the mainstream of scientific opinion. Most doctors and researchers believed that the brain was not cellular in construction but rather consisted of a single, continuous fabric of nerve fibers. And even among those who shared Freud's view that the brain was made of cells, few paid any attention to what might be going on in the spaces between those cells.[3]

Engaged to be wed and in need of a more substantial income, Freud soon abandoned his career as a researcher and went into private practice as a psychoanalyst. But subsequent studies bore out his youthful speculations. Armed with ever more powerful microscopes, scientists confirmed the existence of discrete nerve cells. They also discovered that those cells—our neurons—are both like and unlike the other cells in our bodies. Neurons have central cores, or somas, which carry out the functions common to all cells, but they also have two kinds of tentacle-like appendages—axons and dendrites—that transmit and receive electric pulses. When a neuron is active, a pulse flows from the soma to the tip of the axon, where it triggers the release of chemicals called neurotransmitters. The neurotransmitters flow across Freud's contact barrier—the synapse, we now call it—and attach themselves to a dendrite of a neighboring neuron, triggering (or suppressing) a new electric pulse in that cell. It's through the flow of neurotransmitters across synapses that neurons communicate with one another, directing the transmission of electrical signals along complex cellular pathways. Thoughts, memories, emotions—all emerge from the electrochemical interactions of neurons, mediated by synapses.

During the twentieth century, neuroscientists and psychologists also came to more fully appreciate the astounding complexity of the human brain. Inside our skulls, they discovered, are some 100 billion neurons, which take many different shapes and range in length from a few tenths of a millimeter to a few feet.[4] A single neuron typically has many dendrites (though only one axon), and dendrites and axons can have a multitude of branches and synaptic terminals. The average neuron makes about a thousand synaptic connections, and some neurons can make a hundred times that number. The thousands of billions of synapses inside our skulls tie our neurons together into a dense mesh of circuits that, in ways that are still far from understood, give rise to what we think, how we feel, and who we are.

Even as our knowledge of the physical workings of the brain advanced during the last century, one old assumption remained firmly in place: most biologists and neurologists continued to believe, as they had for hundreds of years, that the structure of the adult brain never changed. Our neurons would connect into circuits during childhood, when our brains were malleable, and as we reached maturity the circuitry would become fixed. The brain, in the prevailing view, was something like a concrete structure. After being poured and shaped in our youth, it hardened quickly into its final form. Once we hit our twenties, no new neurons were created, no new circuits forged. We would, of course, continue to store new memories throughout our lives (and lose some old ones), but the only structural change the brain would go through in adulthood was a slow process of decay as the body aged and nerve cells died.

Although the belief in the adult brain's immutability was deeply and widely held, there were a few heretics. A handful of biologists and psychologists saw in the rapidly growing body of brain research indications that even the adult brain was malleable, or "plastic." New neural circuits could form throughout our lives, they suggested, and old ones might grow stronger or weaker or wither away entirely. The British biologist J. Z. Young, in a series of lectures broadcast by the BBC in 1950, argued that the structure of the brain might in fact be in a constant state of flux, adapting to whatever task it's called on to perform. "There is evidence that the cells of our brains literally develop and grow bigger with use, and atrophy or waste away with disuse," he said. "It may be therefore that every action leaves some permanent print upon the nervous tissue."[5]

Young was not the first to propose such an idea. Seventy years earlier, the American psychologist William James had expressed a similar intuition about the brain's adaptability. The "nervous tissue," he wrote in his landmark *Principles of Psychology*, "seems endowed with a very extraordinary degree of plasticity." As with any other physical compound, "either outward forces or inward tensions can, from one hour to another, turn that structure into something different from what it was." James quoted, approvingly, an analogy that the French scientist Léon Dumont had drawn, in an earlier essay about the biological consequences of habit, between the actions of water on land and the effects of experience on the brain: "Flowing water hollows out a channel for itself which grows broader and deeper; and when it later flows again, it follows the path traced by itself before. Just so, the impressions of outer objects fashion for themselves more and more appropriate paths in the nervous system, and these vital paths recur under similar external stimulation, even if they have been interrupted for some time."[6] Freud, too, ended up taking the contrarian position. In "Project for a Scientific Psychology," a manuscript he wrote in 1895 but never published, he argued that the brain, and in particular the contact barriers between neurons, could change in response to a person's experiences.[7]

Such speculations were dismissed, often contemptuously, by most brain scientists and physicians. They remained convinced that the brain's plasticity ended with childhood, that the "vital paths," once laid, could not be widened or narrowed, much less rerouted. They stood with Santiago Ramón y Cajal, the eminent Spanish physician, neuroanatomist, and Nobel laureate, who in 1913 declared, with a tone that left little room for debate, "In the adult [brain] centres, the nerve paths are something fixed, ended, immutable. Everything may die, nothing may be regenerated."[8] In his younger days, Ramón y Cajal had himself expressed doubts about the orthodox view—he had suggested, in 1894, that the "organ of thought is, within certain limits, malleable, and perfectible by well-directed mental exercise"[9]—but in the end he embraced the conventional wisdom and became one of its most eloquent and authoritative defenders.

The conception of the adult brain as an unchanging physical apparatus grew out of, and was buttressed by, an Industrial Age metaphor that represented the brain as a mechanical contraption. Like a steam engine or an electric dynamo, the nervous system was made up of many parts, and each had a specific and set purpose that contributed in some essential way to the successful operation of the whole. The parts could not change, in shape or function, because that would lead, immediately and inexorably, to the breakdown of the machine. Different regions of the brain, and even individual circuits, played precisely defined roles in processing sensory inputs, directing the movements of muscles, and forming memories and thoughts; and those roles, established in childhood, were not susceptible to alteration. When it came to the brain, the child was indeed, as Wordsworth had written, the father to the man.

The mechanical conception of the brain both reflected and refuted the famous theory of dualism that René Descartes had laid out in his *Meditations* of 1641. Descartes claimed that the brain and the mind existed in two separate spheres: one material, one ethereal.

The physical brain, like the rest of the body, was a purely mechanical instrument that, like a clock or a pump, acted according to the movements of its component parts. But the workings of the brain, argued Descartes, did not explain the workings of the conscious mind. As the essence of the self, the mind existed outside of space, beyond the laws of matter. Mind and brain could influence each other (through, as Descartes saw it, some mysterious action of the pineal gland), but they remained entirely separate substances. At a time of rapid scientific advance and social upheaval, Descartes' dualism came as a comfort. Reality had a material side, which was the realm of science, but it also had a spiritual side, which was the realm of theology—and never the twain shall meet.

As reason became the new religion of the Enlightenment, the notion of an immaterial mind lying outside the reach of observation and experiment seemed increasingly tenuous. Scientists rejected the "mind" half of Cartesian dualism even as they embraced Descartes' idea of the brain as a machine. Thought, memory, and emotion, rather than being the emanations of a spirit world, came to be seen as the logical and predetermined outputs of the physical operations of the brain. Consciousness was simply a by-product of those operations. "The word Mind is obsolete," one prominent neurophysiologist ultimately declared.[10] The machine metaphor was extended, and further reinforced, by the arrival of the digital computer—a "thinking machine"—in the middle of the twentieth century. That's when scientists and philosophers began referring to our brain circuits, and even our behavior, as being "hardwired," just like the microscopic circuits etched into the silicon substrate of a computer chip.

As the idea of the unchangeable adult brain hardened into dogma, it turned into a kind of "neurological nihilism," according to the research psychiatrist Norman Doidge. Because it created "a sense that treatment for many brain problems was ineffective or unwarranted," Doidge explains, it left those with mental illnesses or brain injuries little hope of treatment, much less cure. And as the idea "spread through our culture," it ended up "stunting our overall view of human nature. Since the brain could not change, human nature, which emerges from it, seemed necessarily fixed and unalterable as well."[11] There was no regeneration; there was only decay. We, too, were stuck in the frozen concrete of our brain cells—or at least in the frozen concrete of received wisdom.

IT'S 1968. I'M nine years old, a run-of-the-mill suburban kid playing in a patch of woods near my family's home. Marshall McLuhan and Norman Mailer are on prime-time TV, debating the intellectual and moral implications of what Mailer describes as "man's acceleration into a super-technological world."[12] *2001* is having its first theatrical run, leaving moviegoers befuddled, bemused, or just plain annoyed. And in a quiet laboratory at the University of Wisconsin in Madison, Michael Merzenich is cutting a hole in a monkey's skull.

Twenty-six years old, Merzenich has just received a doctorate in physiology from Johns Hopkins, where he studied under Vernon Mountcastle, a pioneering neuroscientist. He has come to Wisconsin to do postdoctoral research in brain mapping. It's been known for years that every area of a person's body is represented by a corresponding area in the cerebral cortex, the brain's wrinkled outer layer. When certain nerve cells in the skin are stimulated—by being touched or pinched, say—they send an electric pulse through the spinal cord to a particular cluster of neurons in the cortex, which translates the touch or the pinch into a conscious sensation. In the 1930s, the Canadian neurosurgeon Wilder Penfield had used electrical probes to draw the first sensory maps of people's brains. But Penfield's probes were crude instruments, and his maps, while groundbreaking in their time, lacked precision. Merzenich is using a new kind of probe, the hair-thin microelectrode, to create much finer maps that will, he hopes, provide new insight into the brain's structure.

Once he has removed a piece of the monkey's skull and exposed a small portion of its brain, he threads a microelectrode into the

area of the cortex that registers sensations from one of the animal's
hands. He begins tapping that hand in different places until the neuron beside the tip of the electrode fires. After methodically inserting
and reinserting the electrode thousands of times over the course of
a few days, he ends up with a "micromap" showing in minute detail,
down to the individual nerve cell, how the monkey's brain processes
what its hand feels. He repeats the painstaking exercise with five
more monkeys.

Merzenich proceeds to the second stage of his experiment. Using
a scalpel, he makes incisions in the hands of the animals, severing
the sensory nerve. He wants to find out how the brain reacts when a
peripheral nerve system is damaged and then allowed to heal. What
he discovers astounds him. The nerves in the monkeys' hands grow
back in a haphazard fashion, as expected, and their brains, also as
expected, become confused. When, for example, Merzenich touches
the lower joint of a finger on one monkey's hand, the monkey's brain
tells the animal that the sensation is coming from the tip of the finger. The signals have been crossed, the brain map scrambled. But
when Merzenich conducts the same sensory tests a few months later,
he finds that the mental confusion has been cleared up. What the
monkeys' brains tell them is happening to their hands now matches
what's really happening. The brains, Merzenich realizes, have reorganized themselves. The animals' neural pathways have woven themselves into a new map that corresponds to the new arrangement of
nerves in their hands.

At first, he can't believe what he's seen. Like every other neuroscientist, he's been taught that the structure of the adult brain is fixed.
Yet in his lab he has just seen the brains of six monkeys undergo
rapid and extensive restructuring at the cellular level. "I knew it was
astounding reorganization, but I couldn't explain it," Merzenich will
later recall. "Looking back on it, I realized that I had seen evidence
of neuroplasticity. But I didn't know it at the time. I simply didn't
know what I was seeing. And besides, in mainstream neuroscience,
nobody would believe that plasticity was occurring on this scale."[13]

Merzenich publishes the results of his experiment in an academic journal.[14] Nobody pays much heed. But he knows he's onto something, and over the course of the next three decades he conducts many more tests on many more monkeys, all of which point to the existence of broad plasticity in the brains of mature primates. In a 1983 paper documenting one of the experiments, Merzenich declares flatly, "These results are completely contrary to a view of sensory systems as consisting of a series of hardwired machines."[15] At first dismissed, Merzenich's meticulous work finally begins to receive serious notice in the neurological community. It ends up setting off a wholesale reevaluation of accepted theories about how our brains work. Researchers uncover a trail of experiments, dating back to the days of William James and Sigmund Freud, that record examples of plasticity. Long ignored, the old research is now taken seriously.

As brain science continues to advance, the evidence for plasticity strengthens. Using sensitive new brain-scanning equipment, as well as microelectrodes and other probes, neuroscientists conduct more experiments, not only on lab animals but on people. All of them confirm Merzenich's discovery. They also reveal something more: The brain's plasticity is not limited to the somatosensory cortex, the area that governs our sense of touch. It's universal. Virtually all of our neural circuits—whether they're involved in feeling, seeing, hearing, moving, thinking, learning, perceiving, or remembering— are subject to change. The received wisdom is cast aside.

THE ADULT BRAIN, it turns out, is not just plastic but, as James Olds, a professor of neuroscience who directs the Krasnow Institute for Advanced Study at George Mason University, puts it, "very plastic."[16] Or, as Merzenich himself says, "massively plastic."[17] The plasticity diminishes as we get older—brains do get stuck in their ways—but it never goes away. Our neurons are always breaking old connections and forming new ones, and brand-new nerve cells are always being created. "The brain," observes Olds, "has the ability to reprogram itself on the fly, altering the way it functions."

We don't yet know all the details of how the brain reprograms itself, but it has become clear that, as Freud proposed, the secret lies mainly in the rich chemical broth of our synapses. What goes on in the microscopic spaces between our neurons is exceedingly complicated, but in simple terms it involves various chemical reactions that register and record experiences in neural pathways. Every time we perform a task or experience a sensation, whether physical or mental, a set of neurons in our brains is activated. If they're in proximity, these neurons join together through the exchange of synaptic neurotransmitters like the amino acid glutamate.[18] As the same experience is repeated, the synaptic links between the neurons grow stronger and more plentiful through both physiological changes, such as the release of higher concentrations of neurotransmitters, and anatomical ones, such as the generation of new neurons or the growth of new synaptic terminals on existing axons and dendrites. Synaptic links can also weaken in response to experiences, again as a result of physiological and anatomical alterations. What we learn as we live is embedded in the ever-changing cellular connections inside our heads. The chains of linked neurons form our minds' true "vital paths." Today, scientists sum up the essential dynamic of neuroplasticity with a saying known as Hebb's rule: "Cells that fire together wire together."

One of the simplest yet most powerful demonstrations of how synaptic connections change came in a series of experiments that the biologist Eric Kandel performed in the early 1970s on a type of large sea slug called *Aplysia*. (Sea creatures make particularly good subjects for neurological tests because they tend to have simple nervous systems and large nerve cells.) Kandel, who would earn a Nobel Prize for his work, found that if you touch a slug's gill, even very lightly, the gill will immediately and reflexively recoil. But if you touch the gill repeatedly, without causing any harm to the animal, the recoiling instinct will steadily diminish. The slug will become habituated to the touch and learn to ignore it. By monitoring slugs' nervous systems, Kandel discovered that "this learned change in behavior was paralleled by a progressive weakening of the synaptic connections" between the sensory neurons that "feel" the touch and the motor neurons that tell the gill to retract. In a slug's ordinary state, about ninety percent of the sensory neurons in its gill have connections to motor neurons. But after its gill is touched just forty times, only ten percent of the sensory cells maintain links to the motor cells. The research "showed dramatically," Kandel wrote, that "synapses can undergo large and enduring changes in strength after only a relatively small amount of training."[19]

The plasticity of our synapses brings into harmony two philosophies of the mind that have for centuries stood in conflict: empiricism and rationalism. In the view of empiricists, like John Locke, the mind we are born with is a blank slate, a "tabula rasa." What we know comes entirely through our experiences, through what we learn as we live. To put it into more familiar terms, we are products of nurture, not nature. In the view of rationalists, like Immanuel Kant, we are born with built-in mental "templates" that determine how we perceive and make sense of the world. All our experiences are filtered through these inborn templates. Nature predominates.

The *Aplysia* experiments revealed, as Kandel reports, "that both views had merit—in fact they complemented each other." Our genes "specify" many of "the connections among neurons—that is, which neurons form synaptic connections with which other neurons and when." Those genetically determined connections form Kant's innate templates, the basic architecture of the brain. But our experiences regulate the strength, or "long-term effectiveness," of the connections, allowing, as Locke had argued, the ongoing reshaping of the mind and "the expression of new patterns of behavior."[20] The opposing philosophies of the empiricist and the rationalist find their common ground in the synapse. The New York University neuroscientist Joseph LeDoux explains in his book *Synaptic Self* that nature and nurture "actually speak the same language. They both ultimately achieve their mental and behavioral effects by shaping the synaptic organization of the brain."[21]

The brain is not the machine we once thought it to be. Though different regions are associated with different mental functions, the cellular components do not form permanent structures or play rigid roles. They're flexible. They change with experience, circumstance, and need. Some of the most extensive and remarkable changes take place in response to damage to the nervous system. Experiments show, for instance, that if a person is struck blind, the part of the brain that had been dedicated to processing visual stimuli—the visual cortex—doesn't just go dark. It is quickly taken over by circuits used for audio processing. And if the person learns to read Braille, the visual cortex will be redeployed for processing information delivered through the sense of touch.[22] "Neurons seem to 'want' to receive input," explains Nancy Kanwisher of MIT's McGovern Institute for Brain Research: "When their usual input disappears, they start responding to the next best thing."[23] Thanks to the ready adaptability of neurons, the senses of hearing and touch can grow sharper to mitigate the effects of the loss of sight. Similar alterations happen in the brains of people who go deaf: their other senses strengthen to help make up for the loss of hearing. The area in the brain that processes peripheral vision, for example, grows larger, enabling them to see what they once would have heard.

Tests on people who have lost arms or legs in accidents also reveal how extensively the brain can reorganize itself. The areas in the victims' brains that had registered sensations in their lost limbs are quickly taken over by circuits that register sensations from other parts of their bodies. In studying a teenage boy who had lost his left arm in a car crash, the neurologist V. S. Ramachandran, who heads the Center for Brain and Cognition at the University of California at San Diego, discovered that when he had the young man close his eyes and then touched different parts of his face, the patient believed that it was his missing arm that was being touched. At one point, Ramachandran brushed a spot beneath the boy's nose and asked, "Where do you feel that?" The boy replied, "On my left pinky. It tingles." The boy's brain map was in the process of being reorganized, the neurons redeployed for new uses.[24] As a result of such experiments, it's now believed that the sensations of a "phantom limb" felt by amputees are largely the result of neuroplastic changes in the brain.

Our expanding understanding of the brain's adaptability has led to the development of new therapies for conditions that used to be considered untreatable.[25] Doidge, in his 2007 book *The Brain That Changes Itself,* tells the story of a man named Michael Bernstein who suffered a severe stroke when he was fifty-four, damaging an area in the right half of his brain that regulated movement in the left side of his body. Through a traditional program of physical therapy, he recovered some of his motor skills, but his left hand remained crippled and he had to use a cane to walk. Until recently, that would have been the end of the story. But Bernstein enrolled in a program of experimental therapy, run at the University of Alabama by a pioneering neuroplasticity researcher named Edward Taub. For as many as eight hours a day, six days a week, Bernstein used his left hand and his left leg to perform routine tasks over and over again. One day he might wash the pane of a window. The next day he might trace the letters of the alphabet. The repeated actions were a means of coaxing his neurons and synapses to form new circuits that would take over the functions once carried out by the circuits in the damaged area in his brain. In a matter of weeks, he regained nearly all of the movement in his hand and his leg, allowing him to return to his everyday routines and throw away his cane. Many of Taub's other patients have experienced similarly strong recoveries.

Much of the early evidence of neuroplasticity came through the study of the brain's reaction to injuries, whether the severing of the nerves in the hands of Merzenich's monkeys or the loss of sight, hearing, or a limb by human beings. That led some scientists to wonder whether the malleability of the adult brain might be limited to extreme situations. Perhaps, they theorized, plasticity is essentially a healing mechanism, triggered by trauma to the brain or the sensory organs. Further experiments have shown that that's not the case. Extensive, perpetual plasticity has been documented in healthy, normally functioning nervous systems, leading neuroscientists to conclude that our brains are always in flux, adapting to even small shifts in our circumstances and behavior. "We have learned that neuroplasticity is not only possible but that it is constantly in action," writes Mark Hallett, head of the Medical Neurology Branch of the National Institutes of Health. "That is the way we adapt to changing conditions, the way we learn new facts, and the way we develop new skills."[26]

"Plasticity," says Alvaro Pascual-Leone, a top neurology researcher at Harvard Medical School, is "the normal ongoing state of the nervous system throughout the life span." Our brains are constantly changing in response to our experiences and our behavior, reworking their circuitry with "each sensory input, motor act, association, reward signal, action plan, or [shift of] awareness." Neuroplasticity, argues Pascual-Leone, is one of the most important products of evolution, a trait that enables the nervous system "to escape the restrictions of its own genome and thus adapt to environmental pressures, physiologic changes, and experiences."[27] The genius of our brain's construction is not that it contains a lot of hardwiring but that it doesn't. Natural selection, writes the philosopher David Buller in *Adapting Minds*, his critique of evolutionary psychology, "has not designed a brain that consists of numerous prefabricated adaptations" but rather one that is able "to adapt to local environmental demands throughout the lifetime of an individual, and sometimes within a period of days, by forming specialized structures to deal with those demands."[28] Evolution has given us a brain that can literally change its mind—over and over again.

Our ways of thinking, perceiving, and acting, we now know, are not entirely determined by our genes. Nor are they entirely determined by our childhood experiences. We change them through the way we live—and, as Nietzsche sensed, through the tools we use. Years before Edward Taub opened his rehabilitation clinic in Alabama, he conducted a famous experiment on a group of right-handed violinists. Using a machine that monitors neural activity, he measured the areas of their sensory cortex that processed signals from their left hands, the hands they used to finger the strings of their instruments. He also measured the same cortical areas in a group of right-handed volunteers who had never played a musical instrument. He found that the brain areas of the violinists were significantly larger than those of the nonmusicians. He then measured the size of the cortical areas that processed sensations from the subjects' right hands. Here, he found no differences between the musicians and the nonmusicians. Playing a violin, a musical tool, had resulted in substantial physical changes in the brain. That was true even for the musicians who had first taken up their instruments as adults.

When scientists have trained primates and other animals to use simple tools, they've discovered just how profoundly the brain can be influenced by technology. Monkeys, for instance, were taught how to use rakes and pliers to take hold of pieces of food that would otherwise have been out of reach. When researchers monitored the animals' neural activity throughout the course of the training, they found significant growth in the visual and motor areas involved in controlling the hands that held the tools. But they discovered something even more striking as well: the rakes and pliers actually came to be incorporated into the brain maps of the animals' hands. The tools, so far as the animals' brains were concerned, had become part of their bodies. As the researchers who conducted the experiment with the pliers reported, the monkeys' brains began to act "as if the pliers were now the hand fingers."[29]

It's not just repeated physical actions that can rewire our brains. Purely mental activity can also alter our neural circuitry, sometimes in far-reaching ways. In the late 1990s, a group of British researchers scanned the brains of sixteen London cab drivers who had between two and forty-two years of experience behind the wheel. When they compared the scans with those of a control group, they found that the taxi drivers' posterior hippocampus, a part of the brain that plays a key role in storing and manipulating spatial representations of a person's surroundings, was much larger than normal. Moreover, the longer a cab driver had been on the job, the larger his posterior hippocampus tended to be. The researchers also discovered that a portion of the drivers' anterior hippocampus was smaller than average, apparently a result of the need to accommodate the enlargement of the posterior area. Further tests indicated that the shrinking of the anterior hippocampus might have reduced the cabbies' aptitude for certain other memorization tasks. The constant spatial processing required to navigate London's intricate road system, the researchers concluded, is "associated with a relative redistribution of gray matter in the hippocampus."[30]

Another experiment, conducted by Pascual-Leone when he was a researcher at the National Institutes of Health, provides even more remarkable evidence of the way our patterns of thought affect the anatomy of our brains. Pascual-Leone recruited people who had no experience playing a piano, and he taught them how to play a simple melody consisting of a short series of notes. He then split the participants into two groups. He had the members of one group practice the melody on a keyboard for two hours a day over the next five days. He had the members of the other group sit in front of a keyboard for the same amount of time but only imagine playing the song—without ever touching the keys. Using a technique called transcranial magnetic stimulation, or TMS, Pascual-Leone mapped the brain activity of all the participants before, during, and after the test. He found that the people who had only imagined playing the notes exhibited precisely the same changes in their brains as those who had actually pressed the keys.[31] Their brains had changed in response to actions that took place purely in their imagination—in response, that is, to their thoughts. Descartes may have been wrong about dualism, but he appears to have been correct in believing that our thoughts can exert a physical influence on, or at least cause a physical reaction in, our brains. We become, neurologically, what we think.

MICHAEL GREENBERG, IN a 2008 essay in the *New York Review of Books*, found the poetry in neuroplasticity. He observed that our neurological system, "with its branches and transmitters and ingeniously spanned gaps, has an improvised quality that seems to mirror the unpredictability of thought itself." It's "an ephemeral place that changes as our experience changes."[32] There are many reasons to be grateful that our mental hardware is able to adapt so readily to experience, that even old brains can be taught new tricks. The brain's adaptability hasn't just led to new treatments, and new hope, for those suffering from brain injury or illness. It provides all of us with a mental flexibility, an intellectual litheness, that allows us to adapt to new situations, learn new skills, and in general expand our horizons.

But the news is not all good. Although neuroplasticity provides an escape from genetic determinism, a loophole for free thought and free will, it also imposes its own form of determinism on our behavior. As particular circuits in our brain strengthen through the repetition of a physical or mental activity, they begin to transform that activity into a habit. The paradox of neuroplasticity, observes Doidge, is that, for all the mental flexibility it grants us, it can end up locking us into "rigid behaviors."[33] The chemically triggered synapses that link our neurons program us, in effect, to want to keep exercising the circuits they've formed. Once we've wired new circuitry in our brain, Doidge writes, "we long to keep it activated."[34] That's the way the brain fine-tunes its operations. Routine activities are carried out ever more quickly and efficiently, while unused circuits are pruned away.

Plastic does not mean elastic, in other words. Our neural loops don't snap back to their former state the way a rubber band does; they hold onto their changed state. And nothing says the new state has to be a desirable one. Bad habits can be ingrained in our neurons as easily as good ones. Pascual-Leone observes that "plastic changes may not necessarily represent a behavioral gain for a given subject." In addition to being "the mechanism for development and learning," plasticity can be "a cause of pathology."[35]

It comes as no surprise that neuroplasticity has been linked to mental afflictions ranging from depression to obsessive-compulsive disorder to tinnitus. The more a sufferer concentrates on his symptoms, the deeper those symptoms are etched into his neural circuits. In the worst cases, the mind essentially trains itself to be sick. Many addictions, too, are reinforced by the strengthening of plastic pathways in the brain. Even very small doses of addictive drugs can dramatically alter the flow of neurotransmitters in a person's synapses, resulting in long-lasting alterations in brain circuitry and function. In some cases, the buildup of certain kinds of neurotransmitters, such as dopamine, a pleasure-producing cousin to adrenaline, seems to actually trigger the turning on or off of particular genes, bringing even stronger cravings for the drug. The vital paths turn deadly.

The potential for unwelcome neuroplastic adaptations also exists in the everyday, normal functioning of our minds. Experiments show that just as the brain can build new or stronger circuits through physical or mental practice, those circuits can weaken or dissolve with neglect. "If we stop exercising our mental skills," writes Doidge, "we do not just forget them: the brain map space for those skills is turned over to the skills we practice instead."[36] Jeffrey Schwartz, a professor of psychiatry at UCLA's medical school, terms this process "survival of the busiest."[37] The mental skills we sacrifice may be as valuable, or even more valuable, than the ones we gain. When it comes to the *quality* of our thought, our neurons and synapses are entirely indifferent. The possibility of intellectual decay is inherent in the malleability of our brains.

That doesn't mean that we can't, with concerted effort, once again redirect our neural signals and rebuild the skills we've lost. What it does mean is that the vital paths in our brains become, as Monsieur Dumont understood, the paths of least resistance. They are the paths that most of us will take most of the time, and the farther we proceed down them, the more difficult it becomes to turn back.

a digression

---

on what the brain thinks about when it
thinks about itself

THE FUNCTION OF the brain, Aristotle believed, was to keep the body from overheating. A "compound of earth and water," brain matter "tempers the heat and seething of the heart," he wrote in *The Parts of Animals*, a treatise on anatomy and physiology. Blood rises from the "fiery" region of the chest until it reaches the head, where the brain reduces its temperature "to moderation." The cooled blood then flows back down through the rest of the body. The process, suggested Aristotle, was akin to that which "occurs in the production of showers. For when vapor steams up from the earth under the influence of heat and is carried into the upper regions, so soon as it reaches the cold air that is above the earth, it condenses again into water owing to the refrigeration, and falls back to the earth as rain." The reason man has "the largest brain in proportion to his size" is that "the region of the heart and of the lung is hotter and richer in blood in man than in any other animal." It seemed obvious to Aristotle that the brain could not possibly be "the organ of sensation," as Hippocrates and others had conjectured, since "when it is touched, no sensation is produced." In its insensibility, "it resembles," he wrote, "the blood of animals and their excrement."[1]

It's easy, today, to chuckle at Aristotle's error. But it's also easy to understand how the great philosopher was led so far astray. The brain, packed neatly into the bone-crate of the skull, gives us no sensory signal of its existence. We feel our heart beat, our lungs expand, our stomach churn—but our brain, lacking motility and having no sensory nerve endings, remains imperceptible to us. The source of consciousness lies beyond the grasp of consciousness. Physicians and philosophers, from classical times through the Enlightenment, had to deduce the brain's function by examining and dissecting the clumps of grayish tissue they lifted from the skulls of corpses and other dead animals. What they saw usually reflected their assumptions about human nature or, more generally, the nature of the cosmos. They would, as Robert Martensen describes in *The Brain Takes Shape*, fit the visible structure of the brain into their preferred metaphysical metaphor, arranging the organ's physical parts "so as to portray likeness in their own terms."[2]

Writing nearly two thousand years after Aristotle, Descartes conjured up another watery metaphor to explain the brain's function. To him, the brain was a component in an elaborate hydraulic "machine" whose workings resembled those of "fountains in the royal gardens." The heart would pump blood to the brain, where, in the pineal gland, it would be transformed, by means of pressure and heat, into "animal spirits," which then would travel through "the pipes" of the nerves. The brain's "cavities and pores" served as "apertures" regulating the flow of the animal spirits throughout the rest of the body.[3] Descartes' explanation of the brain's role fit neatly into his mechanistic cosmology, in which, as Martensen writes, "*all* bodies operated dynamically according to optical and geometric properties" within self-contained systems.[4]

Our modern microscopes, scanners, and sensors have disabused us of most of the old fanciful notions about the brain's function. But the brain's strangely remote quality—the way it seems both part of us and apart from us—still influences our perceptions in subtle ways. We have a sense that our brain exists in a state of splendid isolation, that its fundamental nature is impervious to the vagaries of our day-to-day lives. While we know that our brain is an exquisitely sensitive monitor of experience, we want to believe that it lies beyond the influence of experience. We want to believe that the impressions our brain records as sensations and stores as memories leave no physical imprint on its own structure. To believe otherwise would, we feel, call into question the integrity of the self.

That was certainly how I felt when I began to worry that my use of the Internet might be changing the way my brain was processing information. I resisted the idea at first. It seemed ludicrous to think that fiddling with a computer, a mere tool, could alter in any deep or lasting way what was going on inside my head. But I was wrong. As neuroscientists have discovered, the brain—and the mind to which it gives rise—is forever a work in progress. That's true not just for each of us as individuals. It's true for all of us as a species.

Three

TOOLS OF THE MIND

A child takes a crayon from a box and scribbles a yellow circle in the corner of a sheet of paper: this is the sun. She takes another crayon and draws a green squiggle through the center of the page: this is the horizon. Cutting through the horizon she draws two brown lines that come together in a jagged peak: this is a mountain. Next to the mountain, she draws a lopsided black rectangle topped by a red triangle: this is her house. The child gets older, goes to school, and in her classroom she traces on a page, from memory, an outline of the shape of her country. She divides it, roughly, into a set of shapes that represent the states. And inside one of the states she draws a five-pointed star to mark the town she lives in. The child grows up. She trains to be a surveyor. She buys a set of fine instruments and uses them to measure the boundaries and contours of a property. With the information, she draws a precise plot of the land, which is then made into a blueprint for others to use.

Our intellectual maturation as individuals can be traced through the way we draw pictures, or maps, of our surroundings. We begin with primitive, literal renderings of the features of the land we see around us, and we advance to ever more accurate, and more abstract, representations of geographic and topographic space. We progress, in other words, from drawing what we see to drawing what we know. Vincent Virga, an expert on cartography affiliated with the Library of Congress, has observed that the stages in the development of our mapmaking skills closely parallel the general stages of childhood cognitive development delineated by the twentieth-century Swiss psychologist Jean Piaget. We progress from the infant's egocentric, purely sensory perception of the world to the young adult's more abstract and objective analysis of experience. "First," writes Virga, in describing how children's drawings of maps advance, "perceptions and representational abilities are not matched; only the simplest topographical relationships are presented, without regard for perspective or distances. Then an intellectual 'realism' evolves, one that depicts everything known with burgeoning proportional relationships. And finally, a visual 'realism' appears, [employing] scientific calculations to achieve it."[1]

As we go through this process of intellectual maturation, we are also acting out the entire history of mapmaking. Mankind's first maps, scratched in the dirt with a stick or carved into a stone with another stone, were as rudimentary as the scribbles of toddlers. Eventually the drawings became more realistic, outlining the actual proportions of a space, a space that often extended well beyond what could be seen with the eye. As more time passed, the realism became scientific in both its precision and its abstraction. The mapmaker began to use sophisticated tools like the direction-finding compass and the angle-measuring theodolite and to rely on mathematical reckonings and formulas. Eventually, in a further intellectual leap, maps came to be used not only to represent vast regions of the earth or heavens in minute detail, but to express ideas—a plan of battle, an analysis of the spread of an epidemic, a forecast of population growth. "The intellectual process of transforming experience *in* space to abstraction *of* space is a revolution in modes of thinking," writes Virga.[2]

The historical advances in cartography didn't simply mirror the development of the human mind. They helped propel and guide the very intellectual advances that they documented. The map is a medium that not only stores and transmits information but also embodies a particular mode of seeing and thinking. As mapmaking progressed, the spread of maps also disseminated the mapmaker's distinctive way of perceiving and making sense of the world. The more frequently and intensively people used maps, the more their minds came to understand reality in the maps' terms. The influence of maps went far beyond their practical employment in establishing property boundaries and charting routes. "The use of a reduced, substitute space for that of reality," explains the cartographic historian Arthur Robinson, "is an impressive act in itself." But what's even more impressive is how the map "advanced the evolution of abstract thinking" throughout society. "The combination of the reduction of reality and the construct of an analogical space is an attainment in abstract thinking of a very high order indeed," writes Robinson, "for it enables one to discover structures that would remain unknown if not mapped."[3] The technology of the map gave to man a new and more comprehending mind, better able to understand the unseen forces that shape his surroundings and his existence.

What the map did for space—translate a natural phenomenon into an artificial and intellectual conception of that phenomenon—another technology, the mechanical clock, did for time. For most of human history, people experienced time as a continuous, cyclical flow. To the extent that time was "kept," the keeping was done by instruments that emphasized this natural process: sundials around which shadows would move, hourglasses down which sand would pour, clepsydras through which water would stream. There was no particular need to measure time with precision or to break a day up into little pieces. For most people, the movements of the sun, the moon, and the stars provided the only clocks they needed. Life was, in the words of the French medievalist Jacques Le Goff, "dominated by agrarian rhythms, free of haste, careless of exactitude, unconcerned by productivity."[4]

That began to change in the latter half of the Middle Ages. The first people to demand a more precise measurement of time were Christian monks, whose lives revolved around a rigorous schedule of prayer. In the sixth century, Saint Benedict had ordered his followers to hold seven prayer services at specified times during the day. Six hundred years later, the Cistercians gave new emphasis to punctuality, dividing the day into a regimented sequence of activities and viewing any tardiness or other waste of time to be an affront to God. Spurred by the need for temporal exactitude, monks took the lead in pushing forward the technologies of timekeeping. It was in the monastery that the first mechanical clocks were assembled, their movements governed by the swinging of weights, and it was the bells in the church tower that first sounded the hours by which people would come to parcel out their lives.

The desire for accurate timekeeping spread outward from the monastery. The royal and princely courts of Europe, brimming with riches and prizing the latest and most ingenious devices, began to covet clocks and invest in their refinement and manufacture. As people moved from the countryside to the town and started working in markets, mills, and factories rather than fields, their days came to be carved into ever more finely sliced segments, each announced by the tolling of a bell. As David Landes describes it in *Revolution in Time*, his history of timekeeping, "Bells sounded for start of work, meal breaks, end of work, closing of gates, start of market, close of market, assembly, emergencies, council meetings, end of drink service, time for street cleaning, curfew, and so on through an extraordinary variety of special peals in individual towns and cities."[5]

The need for tighter scheduling and synchronization of work, transport, devotion, and even leisure provided the impetus for rapid progress in clock technology. It was no longer enough for every town or parish to follow its own clock. Now, time had to be the same everywhere—or else commerce and industry would falter. Units of time became standardized—seconds, minutes, hours—and clock mechanisms were fine-tuned to measure those units with much greater accuracy. By the fourteenth century, the mechanical clock had become commonplace, a near-universal tool for coordinating the intricate workings of the new urban society. Cities vied with one another to install the most elaborate clocks in the towers of their town halls, churches, or palaces. "No European community," the historian Lynn White has observed, "felt able to hold up its head unless in its midst the planets wheeled in cycles and epicycles, while angels trumpeted, cocks crew, and apostles, kings and prophets marched and countermarched at the booming of the hours."[6]

Clocks didn't just become more accurate and more ornate. They got smaller and cheaper. Advances in miniaturization led to the development of affordable timepieces that could fit into the rooms of people's houses or even be carried on their person. If the proliferation of public clocks changed the way people worked, shopped, played, and otherwise behaved as members of an ever more regulated society, the spread of more personal tools for tracking time—chamber clocks, pocket watches, and, a little later, wristwatches—had more intimate consequences. The personal clock became, as Landes writes, "an ever-visible, ever-audible companion and monitor." By continually reminding its owner of "time used, time spent, time wasted, time lost," it became both "prod and key to personal achievement and productivity." The "personalization" of precisely measured time "was a major stimulus to the individualism that was an ever more salient aspect of Western civilization."[7]

The mechanical clock changed the way we saw ourselves. And like the map, it changed the way we thought. Once the clock had redefined time as a series of units of equal duration, our minds began to stress the methodical mental work of division and measurement. We began to see, in all things and phenomena, the pieces that composed the whole, and then we began to see the pieces of which the pieces were made. Our thinking became Aristotelian in its emphasis on discerning abstract patterns behind the visible surfaces of the material world. The clock played a crucial role in propelling us out of the Middle Ages and into the Renaissance and then the Enlightenment.

In *Technics and Civilization*, his 1934 meditation on the human consequences of technology, Lewis Mumford described how the clock
"helped create the belief in an independent world of mathematically measurable sequences." The "abstract framework of divided
time" became "the point of reference for both action and thought."[8]
Independent of the practical concerns that inspired the timekeeping machine's creation and governed its day-to-day use, the clock's
methodical ticking helped bring into being the scientific mind and
the scientific man.

EVERY TECHNOLOGY IS an expression of human will. Through
our tools, we seek to expand our power and control over our
circumstances—over nature, over time and distance, over one
another. Our technologies can be divided, roughly, into four categories, according to the way they supplement or amplify our native
capacities. One set, which encompasses the plow, the darning
needle, and the fighter jet, extends our physical strength, dexterity, or resilience. A second set, which includes the microscope, the
amplifier, and the Geiger counter, extends the range or sensitivity of
our senses. A third group, spanning such technologies as the reser-
voir, the birth control pill, and the genetically modified corn plant,
enables us to reshape nature to better serve our needs or desires.

The map and the clock belong to the fourth category, which
might best be called, to borrow a term used in slightly different
senses by the social anthropologist Jack Goody and the sociologist Daniel Bell, "intellectual technologies." These include all the
tools we use to extend or support our mental powers—to find and
classify information, to formulate and articulate ideas, to share
know-how and knowledge, to take measurements and perform calculations, to expand the capacity of our memory. The typewriter
is an intellectual technology. So are the abacus and the slide rule,
the sextant and the globe, the book and the newspaper, the school
and the library, the computer and the Internet. Although the use of

any kind of tool can influence our thoughts and perspectives—the plow changed the outlook of the farmer, the microscope opened new worlds of mental exploration for the scientist—it is our intellectual technologies that have the greatest and most lasting power over what and how we think. They are our most intimate tools, the ones we use for self-expression, for shaping personal and public identity, and for cultivating relations with others.

What Nietzsche sensed as he typed his words onto the paper clamped in his writing ball—that the tools we use to write, read, and otherwise manipulate information work on our minds even as our minds work with them—is a central theme of intellectual and cultural history. As the stories of the map and the mechanical clock illustrate, intellectual technologies, when they come into popular use, often promote new ways of thinking or extend to the general population established ways of thinking that had been limited to a small, elite group. Every intellectual technology, to put it another way, embodies an intellectual ethic, a set of assumptions about how the human mind works or should work. The map and the clock shared a similar ethic. Both placed a new stress on measurement and abstraction, on perceiving and defining forms and processes beyond those apparent to the senses.

The intellectual ethic of a technology is rarely recognized by its inventors. They are usually so intent on solving a particular problem or untangling some thorny scientific or engineering dilemma that they don't see the broader implications of their work. The users of the technology are also usually oblivious to its ethic. They, too, are concerned with the practical benefits they gain from employing the tool. Our ancestors didn't develop or use maps in order to enhance their capacity for conceptual thinking or to bring the world's hidden structures to light. Nor did they manufacture mechanical clocks to spur the adoption of a more scientific mode of thinking. Those were by-products of the technologies. But what by-products! Ultimately, it's an invention's intellectual ethic that has the most profound effect on us. The intellectual ethic is the message that a medium or other tool transmits into the minds and culture of its users.

For centuries, historians and philosophers have traced, and debated, technology's role in shaping civilization. Some have made the case for what the sociologist Thorstein Veblen dubbed "technological determinism"; they've argued that technological progress, which they see as an autonomous force outside man's control, has been the primary factor influencing the course of human history. Karl Marx gave voice to this view when he wrote, "The windmill gives you society with the feudal lord; the steam-mill, society with the industrial capitalist."[9] Ralph Waldo Emerson put it more crisply: "Things are in the saddle / And ride mankind."[10] In the most extreme expression of the determinist view, human beings become little more than "the sex organs of the machine world," as McLuhan memorably wrote in the "Gadget Lover" chapter of *Understanding Media*.[11] Our essential role is to produce ever more sophisticated tools—to "fecundate" machines as bees fecundate plants—until technology has developed the capacity to reproduce itself on its own. At that point, we become dispensable.

At the other end of the spectrum are the instrumentalists—the people who, like David Sarnoff, downplay the power of technology, believing tools to be neutral artifacts, entirely subservient to the conscious wishes of their users. Our instruments are the means we use to achieve our ends; they have no ends of their own. Instrumentalism is the most widely held view of technology, not least because it's the view we would prefer to be true. The idea that we're somehow controlled by our tools is anathema to most people. "Technology is technology," declared the media critic James Carey; "it is a means for communication and transportation over space, and nothing more."[12]

The debate between determinists and instrumentalists is an illuminating one. Both sides command strong arguments. If you look at a particular technology at a particular point in time, it certainly appears that, as the instrumentalists claim, our tools are firmly under our control. Every day, each of us makes conscious decisions about which tools we use and how we use them. Societies, too, make deliberate choices about how they deploy different technologies. The Japanese, looking to preserve the traditional samurai culture, effectively banned the use of firearms in their country for two centuries. Some religious communities, such as the Old Order Amish fellowships in North America, shun motor cars and other modern technologies. All countries put legal or other restrictions on the use of certain tools.

But if you take a broader historical or social view, the claims of the determinists gain credibility. Although individuals and communities may make very different decisions about which tools they use, that doesn't mean that as a species we've had much control over the path or pace of technological progress. It strains belief to argue that we "chose" to use maps and clocks (as if we might have chosen not to). It's even harder to accept that we "chose" the myriad side effects of those technologies, many of which, as we've seen, were entirely unanticipated when the technologies came into use. "If the experience of modern society shows us anything," observes the political scientist Langdon Winner, "it is that technologies are not merely aids to human activity, but also powerful forces acting to reshape that activity and its meaning."[13] Though we're rarely conscious of the fact, many of the routines of our lives follow paths laid down by technologies that came into use long before we were born. It's an overstatement to say that technology progresses autonomously—our adoption and use of tools are heavily influenced by economic, political, and demographic considerations—but it isn't an overstatement to say that progress has its own logic, which is not always consistent with the intentions or wishes of the toolmakers and tool users. Sometimes our tools do what we tell them to. Other times, we adapt ourselves to our tools' requirements.

The conflict between the determinists and the instrumentalists will never be resolved. It involves, after all, two radically different views of the nature and destiny of humankind. The debate is as much about faith as it is about reason. But there is one thing that determinists and instrumentalists can agree on: technological advances often mark turning points in history. New tools for hunting and farming brought changes in patterns of population growth, settlement, and labor. New modes of transport led to expansions and realignments of trade and commerce. New weaponry altered the balance of power between states. Other breakthroughs, in fields as various as medicine, metallurgy, and magnetism, changed the way people live in innumerable ways—and continue to do so today. In large measure, civilization has assumed its current form as a result of the technologies people have come to use.

What's been harder to discern is the influence of technologies, particularly intellectual technologies, on the functioning of people's brains. We can see the products of thought—works of art, scientific discoveries, symbols preserved on documents—but not the thought itself. There are plenty of fossilized bodies, but there are no fossilized minds. "Gladly would I unfold in calm degrees a natural history of the intellect," wrote Emerson in 1841, "but what man has yet been able to mark the steps and boundaries of that transparent essence?"[14]

Today, at last, the mists that have obscured the interplay between technology and the mind are beginning to lift. The recent discoveries about neuroplasticity make the essence of the intellect more visible, its steps and boundaries easier to mark. They tell us that the tools man has used to support or extend his nervous system—all those technologies that through history have influenced how we find, store, and interpret information, how we direct our attention and engage our senses, how we remember and how we forget—have shaped the physical structure and workings of the human mind. Their use has strengthened some neural circuits and weakened others, reinforced certain mental traits while leaving others to fade away. Neuroplasticity provides the missing link to our understanding of how informational media and other intellectual technologies have exerted their influence over the development of civilization and helped to guide, at a biological level, the history of human consciousness.

We know that the basic form of the human brain hasn't changed much in the last forty thousand years.[15] Evolution at the genetic level proceeds with exquisite slowness, at least when gauged by man's conception of time. But we also know that the ways human beings think and act have changed almost beyond recognition through those millennia. As H. G. Wells observed of mankind in his 1938 book *World Brain*, "His social life, his habits, have changed completely, have even undergone reversion and reversal, while his heredity seems to have changed very little if at all, since the late Stone Age."[16] Our new knowledge of neuroplasticity untangles this conundrum. Between the intellectual and behavioral guardrails set by our genetic code, the road is wide, and we hold the steering wheel. Through what we do and how we do it—moment by moment, day by day, consciously or unconsciously—we alter the chemical flows in our synapses and change our brains. And when we hand down our habits of thought to our children, through the examples we set, the schooling we provide, and the media we use, we hand down as well the modifications in the structure of our brains.

Although the workings of our gray matter still lie beyond the reach of archaeologists' tools, we now know not only that it is probable that the use of intellectual technologies shaped and reshaped the circuitry in our heads, but that it had to be so. Any repeated experience influences our synapses; the changes wrought by the recurring use of tools that extend or supplement our nervous systems should be particularly pronounced. And even though we can't document, at a physical level, the changes in thinking that happened in the distant past, we can use proxies in the present. We see, for example, direct evidence of the ongoing process of mental regeneration and degeneration in the brain changes that occur when a blind person learns to read Braille. Braille, after all, is a technology, an informational medium.

Knowing what we do about London cabbies, we can posit that as people became more dependent on maps, rather than their own memories, in navigating their surroundings, they almost certainly experienced both anatomical and functional changes in the hippocampus and other brain areas involved in spatial modeling and memory. The circuitry devoted to maintaining representations of space likely shrank, while areas employed in deciphering complex and abstract visual information likely expanded or strengthened. We also now know that the changes in the brain spurred by map use could be deployed for other purposes, which helps explain how abstract thinking in general could be promoted by the spread of the cartographer's craft.

The process of our mental and social adaptation to new intellectual technologies is reflected in, and reinforced by, the changing metaphors we use to portray and explain the workings of nature. Once maps had become common, people began to picture all sorts of natural and social relationships as cartographic, as a set of fixed, bounded arrangements in real or figurative space. We began to "map" our lives, our social spheres, even our ideas. Under the sway of the mechanical clock, people began thinking of their brains and their bodies—of the entire universe, in fact—as operating "like clockwork." In the clock's tightly interconnected gears, turning in accord with the laws of physics and forming a long and traceable chain of cause and effect, we found a mechanistic metaphor that seemed to explain the workings of all things, as well as the relations between them. God became the Great Clockmaker. His creation was no longer a mystery to be accepted. It was a puzzle to be worked out. Wrote Descartes in 1646, "Doubtless when the swallows come in spring, they operate like clocks."[17]

THE MAP AND clock changed language indirectly, by suggesting new metaphors to describe natural phenomena. Other intellectual technologies change language more directly, and more deeply, by actually altering the way we speak and listen or read and write. They might enlarge or compress our vocabulary, modify the norms of diction or word order, or encourage either simpler or more complex syntax. Because language is, for human beings, the primary vessel of conscious thought, particularly higher forms of thought, the technologies that restructure language tend to exert the strongest influence over our intellectual lives. As the classical scholar Walter J. Ong put it, "Technologies are not mere exterior aids but also interior transformations of consciousness, and never more than when they affect the word."[18] The history of language is also a history of the mind.

Language itself is not a technology. It's native to our species. Our brains and bodies have evolved to speak and to hear words. A child learns to talk without instruction, as a fledgling bird learns to fly. Because reading and writing have become so central to our identity and culture, it's easy to assume that they, too, are innate talents. But they're not. Reading and writing are unnatural acts, made possible by the purposeful development of the alphabet and many other technologies. Our minds have to be taught how to translate the symbolic characters we see into the language we understand. Reading and writing require schooling and practice, the deliberate shaping of the brain.

Evidence of this shaping process can be seen in many neurological studies. Experiments have revealed that the brains of the literate differ from the brains of the illiterate in many ways—not only in how they understand language but in how they process visual signals, how they reason, and how they form memories. "Learning how to read," reports the Mexican psychologist Feggy Ostrosky-Solís, has been shown to "powerfully shape adult neuropsychological systems."[19] Brain scans have also revealed that people whose written language uses logographic symbols, like the Chinese, develop a mental circuitry for reading that is considerably different from the circuitry found in people whose written language employs a phonetic alphabet. As Tufts University developmental psychologist Maryanne Wolf explains in her book on the neuroscience of reading, *Proust and the Squid*, "Although all reading makes use of some portions of the frontal and temporal lobes for planning and for analyzing sounds and meanings in words, logographic systems appear to activate very distinctive parts of [those] areas, particularly regions involved in motoric memory skills."[20] Differences in brain activity have even been documented among readers of different alphabetic languages. Readers of English, for instance, have been found to draw more heavily on areas of the brain associated with deciphering visual shapes than do readers of Italian. The difference stems, it's believed, from the fact that English words often look very different from the way they sound, whereas in Italian words tend to be spelled exactly as they're spoken.[21]

The earliest examples of reading and writing date back many thousands of years. As long ago as 8000 BC, people were using small clay tokens engraved with simple symbols to keep track of quantities of livestock and other goods. Interpreting even such rudimentary markings required the development of extensive new neural pathways in people's brains, connecting the visual cortex with nearby sense-making areas of the brain. Modern studies show that the neural activity along these pathways doubles or triples when we look at meaningful symbols as opposed to meaningless doodles. As Wolf describes, "Our ancestors could read tokens because their brains were able to connect their basic visual regions to adjacent regions dedicated to more sophisticated visual and conceptual processing."[22] Those connections, which people bequeathed to their children when they taught them to use the tokens, formed the basic wiring for reading.

The technology of writing took an important step forward around the end of the fourth millennium BC. It was then that the Sumerians, living between the Tigris and Euphrates rivers in what is now Iraq, began writing with a system of wedge-shaped symbols, called cuneiform, while a few hundred miles to the west the Egyptians developed increasingly abstract hieroglyphs to represent objects and ideas. Because the cuneiform and hieroglyphic systems incorporated many logosyllabic characters, denoting not just things but also speech sounds, they placed far greater demands on the brain than did the simple accounting tokens. Before readers could interpret the meaning of a character, they had to analyze the character to figure out how it was being used. The Sumerians and the Egyptians had to develop neural circuits that, according to Wolf, literally "crisscrossed" the cortex, linking areas involved not only in seeing and sense-making but in hearing, spatial analysis, and decision making.[23] As these logosyllabic systems expanded to include many hundreds of characters, memorizing and interpreting them became so mentally taxing that their use was probably restricted to an intellectual elite blessed with a lot of time and brain power. For writing technology to progress beyond the Sumerian and Egyptian models, for it to become a tool used by the many rather than the few, it had to get a whole lot simpler.

That didn't happen until fairly recently—around 750 BC—when the Greeks invented the first complete phonetic alphabet. The Greek alphabet had many forerunners, particularly the system of letters developed by the Phoenicians a few centuries earlier, but linguists generally agree that it was the first to include characters representing vowel sounds as well as consonant sounds. The Greeks analyzed all the sounds, or phonemes, used in spoken language, and were able to represent them with just twenty-four characters, making their alphabet a comprehensive and efficient system for writing and reading. The "economy of characters," writes Wolf, reduced "the time and attention needed for rapid recognition" of the symbols and hence required "fewer perceptual and memory resources." Recent brain studies reveal that considerably less of the brain is activated in reading words formed from phonetic letters than in interpreting logograms or other pictorial symbols.[24]

The Greek alphabet became the model for most subsequent Western alphabets, including the Roman alphabet that we still use today. Its arrival marked the start of one of the most far-reaching revolutions in intellectual history: the shift from an oral culture, in which knowledge was exchanged mainly by speaking, to a literary culture, in which writing became the major medium for expressing thought. It was a revolution that would eventually change the lives, and the brains, of nearly everyone on earth, but the transformation was not welcomed by everyone, at least not at first.

Early in the fourth century BC, when the practice of writing was still novel and controversial in Greece, Plato wrote *Phaedrus*, his dialogue about love, beauty, and rhetoric. In the tale, the title character, a citizen of Athens, takes a walk with the great orator Socrates into the countryside, where the two friends sit under a tree beside a stream and have a long and circuitous conversation. They discuss the finer points of speech making, the nature of desire, the varieties of madness, and the journey of the immortal soul, before turning their attention to the written word. "There remains the question," muses Socrates, "of propriety and impropriety in writing."[25] Phaedrus agrees, and Socrates launches into a story about a meeting between the multitalented Egyptian god Theuth, whose many inventions included the alphabet, and one of the kings of Egypt, Thamus.

Theuth describes the art of writing to Thamus and argues that the Egyptians should be allowed to share in its blessings. It will, he says, "make the people of Egypt wiser and improve their memories," for it "provides a recipe for memory and wisdom." Thamus disagrees. He reminds the god that an inventor is not the most reliable judge of the value of his invention: "O man full of arts, to one is it given to create the things of art, and to another to judge what measure of harm and of profit they have for those that shall employ them. And so it is that you, by reason of the tender regard for the writing that is your offspring, have declared the very opposite of its true effect." Should the Egyptians learn to write, Thamus goes on, "it will implant forgetfulness in their souls: they will cease to exercise memory because they rely on that which is written, calling things to remembrance no longer from within themselves, but by means of external marks." The written word is "a recipe not for memory, but for reminder. And it is no true wisdom that you offer your disciples, but only its semblance." Those who rely on reading for their knowledge will "seem to know much, while for the most part they know nothing." They will be "filled, not with wisdom, but with the conceit of wisdom."

Socrates, it's clear, shares Thamus's view. Only "a simple person," he tells Phaedrus, would think that a written account "was at all better than knowledge and recollection of the same matters." Far better than a word written in the "water" of ink is "an intelligent word graven in the soul of the learner" through spoken discourse. Socrates grants that there are practical benefits to capturing one's thoughts in writing—"as memorials against the forgetfulness of old age"—but he argues that a dependence on the technology of the alphabet will alter a person's mind, and not for the better. By substituting outer symbols for inner memories, writing threatens to make us shallower thinkers, he says, preventing us from achieving the intellectual depth that leads to wisdom and true happiness.

Unlike the orator Socrates, Plato was a writer, and while we can assume that he shared Socrates' worry that reading might substitute for remembering, leading to a loss of inner depth, it's also clear that he recognized the advantages that the written word had over the spoken one. In a famous and revealing passage at the end of *The Republic*, a dialogue believed to have been written around the same time as *Phaedrus*, Plato has Socrates go out of his way to attack "poetry," declaring that he would ban poets from his perfect state. Today we think of poetry as being part of literature, a form of writing, but that wasn't the case in Plato's time. Declaimed rather than inscribed, listened to rather than read, poetry represented the ancient tradition of oral expression, which remained central to the Greek educational system, as well as the general Greek culture. Poetry and literature represented opposing ideals of the intellectual life. Plato's argument with the poets, channeled through Socrates' voice, was an argument not against verse but against the oral tradition—the tradition of the bard Homer but also the tradition of Socrates himself—and the ways of thinking it both reflected and encouraged. The "oral state of mind," wrote the British scholar Eric Havelock in *Preface to Plato*, was Plato's "main enemy."[26]

Implicit in Plato's criticism of poetry was, as Havelock, Ong, and other classicists have shown, a defense of the new technology of writing and the state of mind it encouraged in the reader: logical, rigorous, self-reliant. Plato saw the great intellectual benefits that the alphabet could bring to civilization—benefits that were already apparent in his own writing. "Plato's philosophically analytical thought," writes Ong, "was possible only because of the effects that writing was beginning to have on mental processes."[27] In the subtly conflicting views of the value of writing expressed in *Phaedrus* and *The Republic*, we see evidence of the strains created by the transition from an oral to a literary culture. It was, as both Plato and Socrates recognized in their different ways, a shift that was set in motion by the invention of a tool, the alphabet, and that would have profound consequences for our language and our minds.

In a purely oral culture, thinking is governed by the capacity of human memory. Knowledge is what you recall, and what you recall is limited to what you can hold in your mind.[28] Through the millennia of man's preliterate history, language evolved to aid the storage of complex information in individual memory and to make it easy to exchange that information with others through speech. "Serious thought," Ong writes, was by necessity "intertwined with memory systems."[29] Diction and syntax became highly rhythmical, tuned to the ear, and information was encoded in common turns of phrase— what we'd today call clichés—to aid memorization. Knowledge was embedded in "poetry," as Plato defined it, and a specialized class of poet-scholars became the human devices, the flesh-and-blood intellectual technologies, for information storage, retrieval, and transmission. Laws, records, transactions, decisions, traditions—everything that today would be "documented"—in oral cultures had to be, as Havelock says, "composed in formulaic verse" and distributed "by being sung or chanted aloud."[30]

The oral world of our distant ancestors may well have had emotional and intuitive depths that we can no longer appreciate. McLuhan believed that preliterate peoples must have enjoyed a particularly intense "sensuous involvement" with the world. When we learned to read, he argued, we suffered a "considerable detachment from the feelings or emotional involvement that a nonliterate man or society would experience."[31] But intellectually, our ancestors' oral culture was in many ways a shallower one than our own. The written word liberated knowledge from the bounds of individual memory and freed language from the rhythmical and formulaic structures required to support memorization and recitation. It opened to the mind broad new frontiers of thought and expression. "The achievements of the Western world, it is obvious, are testimony to the tremendous values of literacy," McLuhan wrote.[32]

Ong, in his influential 1982 study *Orality and Literacy*, took a similar view. "Oral cultures," he observed, could "produce powerful and beautiful verbal performances of high artistic and human worth, which are no longer even possible once writing has taken possession of the psyche." But literacy "is absolutely necessary for the development not only of science but also of history, philosophy, explicative understanding of literature and of any art, and indeed for the explanation of language (including oral speech) itself."[33] The ability to write is "utterly invaluable and indeed essential for the realization of fuller, interior, human potentials," Ong concluded. "Writing heightens consciousness."[34]

In Plato's time, and for centuries afterward, that heightened consciousness was reserved for an elite. Before the cognitive benefits of the alphabet could spread to the masses, another set of intellectual technologies—those involved in the transcription, production, and distribution of written works—would have to be invented.

Four

THE DEEPENING PAGE

When people first began writing things down, they'd scratch their marks on anything that happened to be lying around—smooth-faced rocks, scraps of wood, strips of bark, bits of cloth, pieces of bone, chunks of broken pottery. Such ephemera were the original media for the written word. They had the advantages of being cheap and plentiful but the disadvantages of being small, irregular in shape, and easily lost, broken, or otherwise damaged. They were suitable for inscriptions and labels, perhaps a brief note or notice, but not much else. No one would think to commit a deep thought or a long argument to a pebble or a potsherd.

The Sumerians were the first to use a specialized medium for writing. They etched their cuneiform into carefully prepared tablets made of clay, an abundant resource in Mesopotamia. They would wash a handful of clay, form it into a thin block, inscribe it with a sharpened reed, and then dry it under the sun or in a kiln. Government records, business correspondence, commercial receipts, and legal agreements were all written on the durable tablets, as were lengthier, more literary works, such as historical and religious stories and accounts of contemporary events. To accommodate the longer pieces of writing, the Sumerians would often number their tablets, creating a sequence of clay "pages" that anticipated the form of the modern book. Clay tablets would continue to be a popular writing medium for centuries, but because preparing, carrying, and storing them were difficult, they tended to be reserved for formal documents written by official scribes. Writing and reading remained arcane talents.

Around 2500 BC, the Egyptians began manufacturing scrolls from the papyrus plants that grew throughout the Nile delta. They would strip fibers from the plants, lay the fibers in a crisscross pattern, and dampen them to release their sap. The resin glued the fibers into a sheet, which was then hammered to form a smooth, white writing surface not all that different from the paper we use today. As many as twenty of the sheets would be glued end to end into long scrolls, and the scrolls, like the earlier clay tablets, would sometimes be arranged in numbered sequences. Flexible, portable, and easy to store, scrolls offered considerable advantages over the much heavier tablets. The Greeks and the Romans adopted scrolls as their primary writing medium, though parchment, made of goat or sheep hide, eventually replaced papyrus as the material of choice in making them.

Scrolls were expensive. Papyrus had to be carted in from Egypt, and turning skins into parchment was a time-consuming job requiring a certain amount of skill. As writing became more common, demand grew for a cheaper option, something that schoolboys could use to take notes and write compositions. That need spurred the development of a new writing device, the wax tablet. It consisted of a simple wooden frame filled with a layer of wax. Letters were scratched into the wax with a new kind of stylus that had, in addition to the sharpened writing tip, a blunt end for scraping the wax clean. Because words could be erased easily from the tablets, students and other writers were able to use them over and over again, making them far more economical than scrolls. Though not a very sophisticated tool, the wax tablet played a major role in turning writing and reading from specialized, formal crafts into casual, everyday activities—for literate citizens, anyway.

The wax tablet was important for another reason. When the ancients wanted an inexpensive way to store or distribute a lengthy text, they would lash a few tablets together with a strip of leather or cloth. These bound tablets, popular in their own right, served as a model for an anonymous Roman artisan who, shortly after the time of Christ, sewed several sheets of parchment between a pair of rigid rectangles of leather to create the first real book. Though a few centuries would pass before the bound book, or codex, supplanted the scroll, the benefits of the technology must have been clear to even its earliest users. Because a scribe could write on both sides of a codex page, a book required much less papyrus or parchment than did a one-sided scroll, reducing the cost of production substantially. Books were also much more compact, making them easier to transport and to conceal. They quickly became the format of choice for publishing early Bibles and other controversial works. Books were easier to navigate too. Finding a particular passage, an awkward task with a long roll of text, became a simple matter of flipping back and forth through a set of pages.

Even as the technology of the book sped ahead, the legacy of the oral world continued to shape the way words on pages were written and read. Silent reading was largely unknown in the ancient world. The new codices, like the tablets and scrolls that preceded them, were almost always read aloud, whether the reader was in a group or alone. In a famous passage in his *Confessions*, Saint Augustine described the surprise he felt when, around the year AD 380, he saw Ambrose, the bishop of Milan, reading silently to himself. "When he read, his eyes scanned the page and his heart explored the meaning, but his voice was silent and his tongue was still," wrote Augustine. "Often, when we came to see him, we found him reading like this in silence, for he never read aloud." Baffled by such peculiar behavior,

Augustine wondered whether Ambrose "needed to spare his voice, which quite easily became hoarse."[1]

It's hard for us to imagine today, but no spaces separated the words in early writing. In the books inked by scribes, words ran together without any break across every line on every page, in what's now referred to as *scriptura continua*. The lack of word separation reflected language's origins in speech. When we talk, we don't insert pauses between each word—long stretches of syllables flow unbroken from our lips. It would never have crossed the minds of the first writers to put blank spaces between words. They were simply transcribing speech, writing what their ears told them to write. (Today, when young children begin to write, they also run their words together. Like the early scribes, they write what they hear.) The scribes didn't pay much attention to the order of the words in a sentence either. In spoken language, meaning had always been conveyed mainly through inflection, the pattern of stresses a speaker places on syllables, and that oral tradition continued to govern writing. In interpreting the writing in books through the early Middle Ages, readers would not have been able to use word order as a signal of meaning. The rules hadn't been invented yet.[2]

The lack of word separation, combined with the absence of word order conventions, placed an "extra cognitive burden" on ancient readers, explains Paul Saenger in *Space between Words*, his history of the scribal book.[3] Readers' eyes had to move slowly and haltingly across the lines of text, pausing frequently and often backing up to the start of a sentence, as their minds struggled to figure out where one word ended and a new one began and what role each word was playing in the meaning of the sentence. Reading was like working out a puzzle. The brain's entire cortex, including the forward areas associated with problem solving and decision making, would have been buzzing with neural activity.

The slow, cognitively intensive parsing of text made the reading of books laborious. It was also the reason no one, other than the odd case like Ambrose, read silently. Sounding out the syllables was crucial to deciphering the writing. Those constraints, which would seem intolerable to us today, didn't matter much in a culture still rooted in orality. "Because those who read relished the mellifluous metrical and accentual patterns of pronounced text," writes Saenger, "the absence of interword space in Greek and Latin was not perceived to be an impediment to effective reading, as it would be to the modern reader, who strives to read swiftly."[4] Besides, most literate Greeks and Romans were more than happy to have their books read to them by slaves.

NOT UNTIL WELL after the collapse of the Roman Empire did the form of written language finally break from the oral tradition and begin to accommodate the unique needs of readers. As the Middle Ages progressed, the number of literate people—cenobites, students, merchants, aristocrats—grew steadily, and the availability of books expanded. Many of the new books were of a technical nature, intended not for leisurely or scholarly reading but for practical reference. People began to want, and to need, to read quickly and privately. Reading was becoming less an act of performance and more a means of personal instruction and improvement. That shift led to the most important transformation of writing since the invention of the phonetic alphabet. By the start of the second millennium, writers had begun to impose rules of word order on their work, fitting words into a predictable, standardized system of syntax. At the same time, beginning in Ireland and England and then spreading throughout the rest of western Europe, scribes started dividing sentences into individual words, separated by spaces. By the thirteenth century, *scriptura continua* was largely obsolete, for Latin texts as well as those written in the vernacular. Punctuation marks, which further eased the work of the reader, began to become common too. Writing, for the first time, was aimed as much at the eye as the ear.

It would be difficult to overstate the significance of these changes.

The emergence of word order standards sparked a revolution in the structure of language—one that, as Saenger notes, "was inherently antithetical to the ancient quest for metrical and rhythmical eloquence."[5] The placing of spaces between words alleviated the cognitive strain involved in deciphering text, making it possible for people to read quickly, silently, and with greater comprehension. Such fluency had to be learned. It required complex changes in the circuitry of the brain, as contemporary studies of young readers reveal. The accomplished reader, Maryanne Wolf explains, develops specialized brain regions geared to the rapid deciphering of text. The areas are wired "to represent the important visual, phonological, and semantic information and to retrieve this information at lightning speed." The visual cortex, for example, develops "a veritable collage" of neuron assemblies dedicated to recognizing, in a matter of milliseconds, "visual images of letters, letter patterns, and words."[6] As the brain becomes more adept at decoding text, turning what had been a demanding problem-solving exercise into a process that is essentially automatic, it can dedicate more resources to the interpretation of meaning. What we today call "deep reading" becomes possible. By "altering the neurophysiological process of reading," word separation "freed the intellectual faculties of the reader," Saenger writes; "even readers of modest intellectual capacity could read more swiftly, and they could understand an increasing number of inherently more difficult texts."[7]

Readers didn't just become more efficient. They also became more attentive. To read a long book silently required an ability to concentrate intently over a long period of time, to "lose oneself" in the pages of a book, as we now say. Developing such mental discipline was not easy. The natural state of the human brain, like that of the brains of most of our relatives in the animal kingdom, is one of distractedness. Our predisposition is to shift our gaze, and hence our attention, from one object to another, to be aware of as much of what's going on around us as possible. Neuroscientists have discovered primitive "bottom-up mechanisms" in our brains that, as the authors of a 2004 article in *Current Biology* put it, "operate on raw sensory input, rapidly and involuntarily shifting attention to salient visual features of potential importance."[8] What draws our attention most of all is any hint of a change in our surroundings. "Our senses are finely attuned to change," explains Maya Pines of the Howard Hughes Medical Institute. "Stationary or unchanging objects become part of the scenery and are mostly unseen." But as soon as "something in the environment changes, we need to take notice because it might mean danger —or opportunity."[9] Our fast-paced, reflexive shifts in focus were once crucial to our survival. They reduced the odds that a predator would take us by surprise or that we'd overlook a nearby source of food. For most of history, the normal path of human thought was anything but linear.

To read a book was to practice an unnatural process of thought, one that demanded sustained, unbroken attention to a single, static object. It required readers to place themselves at what T. S. Eliot, in *Four Quartets*, would call "the still point of the turning world." They had to train their brains to ignore everything else going on around them, to resist the urge to let their focus skip from one sensory cue to another. They had to forge or strengthen the neural links needed to counter their instinctive distractedness, applying greater "top-down control" over their attention.[10] "The ability to focus on a single task, relatively uninterrupted," writes Vaughan Bell, a research psychologist at King's College London, represents a "strange anomaly in the history of our psychological development."[11]

Many people had, of course, cultivated a capacity for sustained attention long before the book or even the alphabet came along. The hunter, the craftsman, the ascetic—all had to train their brains to control and concentrate their attention. What was so remarkable about book reading was that the deep concentration was combined with the highly active and efficient deciphering of text and interpretation of meaning. The reading of a sequence of printed pages was valuable not just for the knowledge readers acquired from the author's words but for the way those words set off intellectual vibrations within their own minds. In the quiet spaces opened up by the prolonged, undistracted reading of a book, people made their own associations, drew their own inferences and analogies, fostered their own ideas. They thought deeply as they read deeply.

Even the earliest silent readers recognized the striking change in their consciousness that took place as they immersed themselves in the pages of a book. The medieval bishop Isaac of Syria described how, whenever he read to himself, "as in a dream, I enter a state when my sense and thoughts are concentrated. Then, when with prolonging of this silence the turmoil of memories is stilled in my heart, ceaseless waves of joy are sent me by inner thoughts, beyond expectation suddenly arising to delight my heart."[12] Reading a book was a meditative act, but it didn't involve a clearing of the mind. It involved a filling, or replenishing, of the mind. Readers disengaged their attention from the outward flow of passing stimuli in order to engage it more deeply with an inward flow of words, ideas, and emotions. That was—and is—the essence of the unique mental process of deep reading. It was the technology of the book that made this "strange anomaly" in our psychological history possible. The brain of the book reader was more than a literate brain. It was a literary brain.

The changes in written language liberated the writer as well as the reader. *Scriptura continua* wasn't just a nuisance to decipher; it was a trial to write. To escape the drudgery, writers would usually dictate their works to a professional scribe. As soon as the introduction of word spaces made writing easier, authors took up pens and began putting their words onto the page themselves, in private. Their works immediately became more personal and more adventurous. They began to give voice to unconventional, skeptical, and even heretical and seditious ideas, pushing the bounds of knowledge and culture. Working alone in his chambers, the Benedictine monk Guibert of Nogent had the confidence to compose unorthodox interpretations of scripture, vivid accounts of his dreams, even erotic poetry—things he would never have written had he been required to dictate them to a scribe. When, late in his life, he lost his sight and had to go back to dictation, he complained of having to write "only by voice, without the hand, without the eyes."[13]

Authors also began to revise and edit their works heavily, something that dictation had often precluded. That, too, altered the form and the content of writing. For the first time, explains Saenger, a writer "could see his manuscript as a whole and by means of cross-references develop internal relationships and eliminate the redundancies common to the dictated literature" of the earlier Middle Ages.[14] The arguments in books became longer and clearer, as well as more complex and more challenging, as writers strived self-consciously to refine their ideas and their logic. By the end of the fourteenth century, written works were often being divided into paragraphs and chapters, and they sometimes included tables of contents to help guide the reader through their increasingly elaborate structures.[15] There had, of course, been sensitive and self-conscious prose and verse stylists in the past, as Plato's dialogues elegantly demonstrate, but the new writing conventions greatly expanded the production of literary works, particularly those composed in the vernacular.

The advances in book technology changed the personal experience of reading and writing. They also had social consequences. The broader culture began to mold itself, in ways both subtle and obvious, around the practice of silent book reading. The nature of education and scholarship changed, as universities began to stress private reading as an essential complement to classroom lectures. Libraries began to play much more central roles in university life and, more generally, in the life of the city. Library architecture evolved too. Private cloisters and carrels, tailored to accommodate vocal reading, were torn out and replaced by large public rooms where students, professors, and other patrons sat together at long tables reading silently to themselves. Reference books such as dictionaries, glossaries, and concordances became important as aids to reading. Copies of the precious texts were often chained to the library reading tables. To fill the increasing demand for books, a publishing industry started to take shape. Book production, long the realm of the religious scribe working in a monastery's scriptorium, started to be centralized in secular workshops, where professional scribes worked for pay under the direction of the owner. A lively market for used books materialized. For the first time in history, books had set prices.[16]

For centuries, the technology of writing had reflected, and reinforced, the intellectual ethic of the oral culture in which it arose. The writing and reading of tablets, scrolls, and early codices had stressed the communal development and propagation of knowledge. Individual creativity had remained subordinate to the needs of the group. Writing had remained more a means of recording than a method of composition. Now, writing began to take on, and to disseminate, a new intellectual ethic: the ethic of the book. The development of knowledge became an increasingly private act, with each reader creating, in his own mind, a personal synthesis of the ideas and information passed down through the writings of other thinkers. The sense of individualism strengthened. "Silent reading," the novelist and historian James Carroll has noted, is "both the sign of and a means to self-awareness, with the knower taking responsibility for what is known."[17] Quiet, solitary research became a prerequisite for intellectual achievement. Originality of thought and creativity of expression became the hallmarks of the model mind. The conflict between the orator Socrates and the writer Plato had at last been decided—in Plato's favor.

But the victory was incomplete. Because handwritten codices remained costly and scarce, the intellectual ethic of the book, and the mind of the deep reader, continued to be restricted to a relatively small group of privileged citizens. The alphabet, a medium of language, had found its own ideal medium in the book, a medium of writing. Books, however, had yet to find their ideal medium—the technology that would allow them to be produced and distributed cheaply, quickly, and in abundance.

SOMETIME AROUND 1445, a German goldsmith named Johannes Gutenberg left Strasbourg, where he had been living for several years, and followed the Rhine River back to the city of his birth, Mainz. He was carrying a secret—a big one. For at least ten years, he had been working covertly on several inventions that he believed would, in combination, form the basis of an altogether new sort of publishing business. He saw an opportunity to automate the production of books and other written works, replacing the venerable scribe with a newfangled printing machine. After securing two sizable loans from Johann Fust, a prosperous neighbor, Gutenberg set up a shop in Mainz, bought some tools and materials, and set to work. Putting his metalworking skills to use, he created small, adjustable molds for casting alphabetical letters of uniform height but varying width out of a molten metal alloy. The cast letters, or movable type, could be arranged quickly into a page of text for printing and then, when the job was done, disassembled and reset for a new page.[18] Gutenberg also developed a refined version of a wooden-screw press, used at the time to crush grapes for wine, that was able to transfer the image of the type onto a sheet of parchment or paper without smudging the letters. And he invented the third critical element of his printing system: an oil-based ink that would adhere to the metal type.

Having built the letterpress, Gutenberg quickly put it to use printing indulgences for the Catholic Church. The job paid well, but it wasn't the work Gutenberg had in mind for his new machine. He had much greater ambitions. Drawing on Fust's funds, he began to prepare his first major work: the magnificent, two-volume edition of the Bible that would come to bear his name. Spanning twelve hundred pages, each composed of two forty-two-line columns, the Gutenberg Bible was printed in a heavy Gothic typeface painstakingly designed to imitate the handwriting of the best German scribes. The Bible, which took at least three years to produce, was Gutenberg's triumph. It was also his undoing. In 1455, having printed just two hundred copies, he ran out of money. Unable to pay the interest on his loans, he was forced to hand his press, type, and ink over to Fust and abandon the printing trade. Fust, who had made his fortune through a successful career as a merchant, proved to be as adept at the business of printing as Gutenberg had been at its mechanics. Together with Peter Schoeffer, one of Gutenberg's more talented employees (and a former scribe himself), Fust set the operation on a profitable course, organizing a sales force and publishing a variety of books that sold widely throughout Germany and France.[19]

Although Gutenberg would not share in its rewards, his letterpress would become one of the most important inventions in history. With remarkable speed, at least by medieval standards, movable-type printing "changed the face and condition of things all over the world," Francis Bacon wrote in his 1620 book *Novum Organum*, "so that no empire or sect or star seems to have exercised a greater power and influence on human affairs."[20] (The only other inventions that Bacon felt had as great an impact as the letterpress were gunpowder and the compass.) By turning a manual craft into a mechanical industry, Gutenberg had changed the economics of printing and publishing. Large editions of perfect copies could be mass-produced quickly by a few workers. Books went from being expensive, scarce commodities to being affordable, plentiful ones.

In 1483, a printing shop in Florence, run by nuns from the Convent of San Jacopo di Ripoli, charged three florins for printing 1,025 copies of a new translation of Plato's *Dialogues*. A scribe would have charged about one florin for copying the work, but he would have produced only a single copy.[21] The steep reduction in the cost of manufacturing books was amplified by the growing use of paper, an invention imported from China, in place of more costly parchment. As book prices fell, demand surged, spurring, in turn, a rapid expansion in supply. New editions flooded the markets of Europe. According to one estimate, the number of books produced in the fifty years following Gutenberg's invention equaled the number produced by European scribes during the preceding thousand years.[22] The sudden proliferation of once-rare books struck people of the time "as sufficiently remarkable to suggest supernatural intervention," reports Elizabeth Eisenstein in *The Printing Press as an Agent of Change*.[23] When Johann Fust carried a large supply of printed books into Paris on an early sales trip, he was reportedly run out of town by the gendarmes on suspicion of being in league with the devil.[24]

Fears of satanic influence quickly dissipated as people rushed to buy and read the inexpensive products of the letterpress. When, in 1501, the Italian printer Aldus Manutius introduced the pocket-sized octavo format, considerably smaller than the traditional folio and quarto, books became even more affordable, portable, and personal. Just as the miniaturization of the clock made everyone a timekeeper, so the miniaturization of the book helped weave book-reading into the fabric of everyday life. It was no longer just scholars and monks who sat reading words in quiet rooms. Even a person of fairly modest means could begin to assemble a library of several volumes, making it possible not only to read broadly but to draw comparisons between different works. "All the world is full of knowing men, of most learned Schoolmasters, and vast Libraries," exclaimed the title character of Rabelais' 1534 best seller *Gargantua*, "and it appears to me as a truth, that neither in Plato's time, nor Cicero's, nor Papinian's, there was ever such conveniency for studying, as we see at this day there is."[25]

A virtuous cycle had been set in motion. The growing availability of books fired the public's desire for literacy, and the expansion of literacy further stimulated the demand for books. The printing industry boomed. By the end of the fifteenth century, nearly 250 towns in Europe had print shops, and some 12 million volumes had already come off their presses. The sixteenth century saw Gutenberg's technology leap from Europe to Asia, the Middle East, and, when the Spanish set up a press in Mexico City in 1539, the Americas. By the start of the seventeenth century, letterpresses were everywhere, producing not only books but newspapers, scientific journals, and a variety of other periodicals. The first great flowering of printed literature arrived, with works by such masters as Shakespeare, Cervantes, Molière, and Milton, not to mention Bacon and Descartes, entering the inventories of booksellers and the libraries of readers.

It wasn't just contemporary works that were coming off the presses. Printers, striving to fill the public's demand for inexpensive reading material, produced large editions of the classics, both in the original Greek and Latin and in translation. Although most of the printers were motivated by the desire to turn an easy profit, the distribution of the older texts helped give intellectual depth and historical continuity to the emerging book-centered culture. As Eisenstein writes, the printer who "duplicated a seemingly antiquated backlist" may have been lining his own pockets, but in the process he gave readers "a richer, more varied diet than had been provided by the scribe."[26]

Along with the high-minded came the low-minded. Tawdry novels, quack theories, gutter journalism, propaganda, and, of course, reams of pornography poured into the marketplace and found eager buyers at every station in society. Priests and politicians began to wonder whether, as England's first official book censor put it in 1660, "more mischief than advantage were not occasion'd to the Christian world by the Invention of Typography."[27] The famed Spanish dramatist Lope de Vega expressed the feelings of many a grandee when, in his 1612 play *All Citizens Are Soldiers*, he wrote:

So many books—so much confusion!
All around us an ocean of print
And most of it covered in froth.[28]

But the froth itself was vital. Far from dampening the intellectual transformation wrought by the printed book, it magnified it. By accelerating the spread of books into popular culture and making them a mainstay of leisure time, the cruder, crasser, and more trifling works also helped spread the book's ethic of deep, attentive reading. "The same silence, solitude, and contemplative attitudes associated formerly with pure spiritual devotion," writes Eisenstein, "also accompanies the perusal of scandal sheets, 'lewd Ballads,' 'merry bookes of Italie,' and other 'corrupted tales in Inke and

Paper.'"[29] Whether a person is immersed in a bodice ripper or a Psalter, the synaptic effects are largely the same.

Not everyone became a book reader, of course. Plenty of people—the poor, the illiterate, the isolated, the incurious—never participated, at least not directly, in Gutenberg's revolution. And even among the most avid of the book-reading public, many of the old oral practices of information exchange remained popular. People continued to chat and to argue, to attend lectures, speeches, debates, and sermons.[30] Such qualifications deserve note—any generalization about the adoption and use of a new technology will be imperfect—but they don't change the fact that the arrival of movable-type printing was a central event in the history of Western culture and the development of the Western mind.

"For the medieval type of brain," writes J. Z. Young, "making true statements depended on fitting sensory experience with the symbols of religion." The letterpress changed that. "As books became common, men could look more directly at each other's observations, with a great increase in the accuracy and content of the information conveyed."[31] Books allowed readers to compare their thoughts and experiences not just with religious precepts, whether embedded in symbols or voiced by the clergy, but with the thoughts and experiences of others.[32] The social and cultural consequences were as widespread as they were profound, ranging from religious and political upheaval to the ascendancy of the scientific method as the central means for defining truth and making sense of existence. What was widely seen as a new "Republic of Letters" came into being, open at least theoretically to anyone able to exercise, as the Harvard historian Robert Darnton puts it, "the two main attributes of citizenship, writing and reading."[33] The literary mind, once confined to the cloisters of the monastery and the towers of the university, had become the general mind. The world, as Bacon recognized, had been remade.

THERE ARE MANY kinds of reading. David Levy, in *Scrolling Forward*, a book about our present-day transition from printed to electronic documents, notes that literate people "read all day long, mostly unconsciously." We glance at road signs, menus, headlines, shopping lists, the labels of products in stores. "These forms of reading," he says, "tend to be shallow and of brief duration." They're the types of reading we share with our distant ancestors who deciphered the marks scratched on pebbles and potsherds. But there are also times, Levy continues, "when we read with greater intensity and duration, when we become absorbed in what we are reading for longer stretches of time. Some of us, indeed, don't just *read* in this way but think of ourselves as *readers*."[34]

Wallace Stevens, in the exquisite couplets of "The House Was Quiet and the World Was Calm," provides a particularly memorable and moving portrayal of the kind of reading Levy is talking about:

The house was quiet and the world was calm.
The reader became the book; and summer night

Was like the conscious being of the book.
The house was quiet and the world was calm.

The words were spoken as if there was no book,
Except that the reader leaned above the page,

Wanted to lean, wanted much most to be
The scholar to whom his book is true, to whom

The summer night is like a perfection of thought.
The house was quiet because it had to be.

The quiet was part of the meaning, part of the mind:
The access of perfection to the page.

Stevens' poem not only describes deep reading. It demands deep reading. The apprehension of the poem requires the mind the poem describes. The "quiet" and the "calm" of the deep reader's attentiveness become "part of the meaning" of the poem, forming the pathway through which "perfection" of thought and expression reaches the page. In the metaphorical "summer night" of the wholly engaged intellect, the writer and the reader merge, together creating and sharing "the conscious being of the book."

Recent research into the neurological effects of deep reading has added a scientific gloss to Stevens' lyric. In one fascinating study, conducted at Washington University's Dynamic Cognition Laboratory and published in the journal *Psychological Science* in 2009, researchers used brain scans to examine what happens inside people's heads as they read stories. They found that "readers mentally simulate each new situation encountered in a narrative. Details about actions and sensation are captured from the text and integrated with personal knowledge from past experiences." The brain regions that are activated often "mirror those involved when people perform, imagine, or observe similar real-world activities." Deep reading, says the study's lead researcher, Nicole Speer, "is by no means a passive exercise."[35] The reader becomes the book.

The bond between book reader and book writer has always been a tightly symbiotic one, a means of intellectual and artistic cross-fertilization. The words of the writer act as a catalyst in the mind of the reader, inspiring new insights, associations, and perceptions, sometimes even epiphanies. And the very existence of the attentive, critical reader provides the spur for the writer's work. It gives the author the confidence to explore new forms of expression, to blaze difficult and demanding paths of thought, to venture into uncharted and sometimes hazardous territory. "All great men have written proudly, nor cared to explain," said Emerson. "They knew that the intelligent reader would come at last, and would thank them."[36]

Our rich literary tradition is unthinkable without the intimate exchanges that take place between reader and writer within the crucible of a book. After Gutenberg's invention, the bounds of language expanded rapidly as writers, competing for the eyes of ever more sophisticated and demanding readers, strived to express ideas and emotions with superior clarity, elegance, and originality. The vocabulary of the English language, once limited to just a few thousand words, expanded to upwards of a million words as books proliferated.[37] Many of the new words encapsulated abstract concepts that simply hadn't existed before. Writers experimented with syntax and diction, opening new pathways of thought and imagination. Readers eagerly traveled down those pathways, becoming adept at following fluid, elaborate, and idiosyncratic prose and verse. The ideas that writers could express and readers could interpret became more complex and subtle, as arguments wound their way linearly across many pages of text. As language expanded, consciousness deepened.

The deepening extended beyond the page. It's no exaggeration to say that the writing and reading of books enhanced and refined people's experience of life and of nature. "The remarkable virtuosity displayed by new literary artists who managed to counterfeit taste, touch, smell, or sound in mere words required a heightened awareness and closer observation of sensory experience that was passed on in turn to the reader," writes Eisenstein. Like painters and composers, writers were able "to alter perception" in a way "that enriched rather than stunted sensuous response to external stimuli, expanded rather than contracted sympathetic response to the varieties of human experience."[38] The words in books didn't just strengthen people's ability to think abstractly; they enriched people's experience of the physical world, the world outside the book.

One of the most important lessons we've learned from the study of neuroplasticity is that the mental capacities, the very neural circuits, we develop for one purpose can be put to other uses as well. As our ancestors imbued their minds with the discipline to follow a line of argument or narrative through a succession of printed pages, they became more contemplative, reflective, and imaginative. "New thought came more readily to a brain that had already learned how to rearrange itself to read," says Maryanne Wolf; "the increasingly sophisticated intellectual skills promoted by reading and writing added to our intellectual repertoire."[39] The quiet of deep reading became, as Stevens understood, "part of the mind."

Books weren't the only reason that human consciousness was transformed during the years following the invention of the letterpress—many other technologies and social and demographic trends played important roles—but books were at the very center of the change. As the book came to be the primary means of exchanging knowledge and insight, its intellectual ethic became the foundation of our culture. The book made possible the delicately nuanced self-knowledge found in Wordsworth's *Prelude* and Emerson's essays and the equally subtle understanding of social and personal relations found in the novels of Austen, Flaubert, and Henry James. Even the great twentieth-century experiments in nonlinear narrative by writers like James Joyce and William Burroughs would have been unthinkable without the artists' presumption of attentive, patient readers. When transcribed to a page, a stream of consciousness becomes literary and linear.

The literary ethic was not only expressed in what we normally think of as literature. It became the ethic of the historian, illuminating works like Gibbon's *Decline and Fall of the Roman Empire*. It became the ethic of the philosopher, informing the ideas of Descartes, Locke, Kant, and Nietzsche. And, crucially, it became the ethic of the scientist. One could argue that the single most influential literary work of the nineteenth century was Darwin's *On the Origin of Species*. In the twentieth century, the literary ethic ran through such diverse books as Einstein's *Relativity*, Keynes's *General Theory of Employment, Interest and Money*, Thomas Kuhn's *Structure of Scientific Revolutions*, and Rachel Carson's *Silent Spring*. None of these momentous intellectual achievements would have been possible without the changes in reading and writing—and in perceiving and thinking—spurred by the efficient reproduction of long forms of writing on printed pages.

LIKE OUR FOREBEARS during the later years of the Middle Ages, we find ourselves today between two technological worlds. After 550 years, the printing press and its products are being pushed from the center of our intellectual life to its edges. The shift began during the middle years of the twentieth century, when we started devoting more and more of our time and attention to the cheap, copious, and endlessly entertaining products of the first wave of electric and electronic media: radio, cinema, phonograph, television. But those technologies were always limited by their inability to transmit the written word. They could displace but not replace the book. Culture's mainstream still ran through the printing press.

Now the mainstream is being diverted, quickly and decisively, into a new channel. The electronic revolution is approaching its culmination as the computer—desktop, laptop, handheld—becomes our constant companion and the Internet becomes our medium of choice for storing, processing, and sharing information in all forms, including text. The new world will remain, of course, a literate world, packed with the familiar symbols of the alphabet. We cannot go back to the lost oral world, any more than we can turn the clock back to a time before the clock existed.[40] "Writing and print and the computer," writes Walter Ong, "are all ways of technologizing the word"; and once technologized, the word cannot be de-technologized.[41] But the world of the screen, as we're already coming to understand, is a very different place from the world of the page. A new intellectual ethic is taking hold. The pathways in our brains are once again being rerouted.

a digression

on lee de forest and his amazing audion

OUR MODERN MEDIA spring from a common source, an invention that is rarely mentioned today but that had as decisive a role in shaping society as the internal combustion engine or the incandescent lightbulb. The invention was called the Audion. It was the first electronic audio amplifier, and the man who created it was Lee de Forest.

Even when judged by the high standards set by America's mad-genius inventors, de Forest was an oddball. Nasty, ill-favored, and generally despised—in high school he was voted "homeliest boy" in his class—he was propelled by an enormous ego and an equally out-sized inferiority complex.[1] When he wasn't marrying or divorcing a wife, alienating a colleague, or leading a business to ruin, he was usually in court defending himself against charges of fraud or patent infringement—or pressing his own suit against one of his many enemies.

De Forest grew up in Alabama, the son of a schoolmaster. After earning a doctorate in engineering from Yale in 1896, he spent a decade fiddling with the latest radio and telegraph technology, desperately seeking the breakthrough that would make his name and fortune. In 1906, his moment arrived. Without quite knowing what he was doing, he took a standard two-pole vacuum tube, which sent an electric current from one wire (the filament) to a second (the plate), and he added a third wire to it, turning the diode into a triode. He found that when he sent a small electric charge into the third wire—the grid—it boosted the strength of the current running between the filament and the plate. The device, he explained in a patent application, could be adapted "for amplifying feeble electric currents."[2]

De Forest's seemingly modest invention turned out to be a world changer. Because it could be used to amplify an electrical signal, it could also be used to amplify audio transmissions sent and received as radio waves. Up to then, radios had been of limited use because their signals faded so quickly. With the Audion to boost the signals, long-distance wireless transmissions became possible, setting the stage for radio broadcasting. The Audion became, as well, a critical component of the new telephone system, enabling people on opposite sides of the country, or the world, to hear each other talk.

De Forest couldn't have known it at the time, but he had inaugurated the age of electronics. Electric currents are, simply put, streams of electrons, and the Audion was the first device that allowed the intensity of those streams to be controlled with precision. As the twentieth century progressed, triode tubes came to form the technological heart of the modern communications, entertainment, and media industries. They could be found in radio transmitters and receivers, in hi-fi sets, in public address systems, in guitar amps. Arrays of tubes also served as the processing units and data storage systems in many early digital computers. The first mainframes often had tens of thousands of them. When, around 1950, vacuum tubes began to be replaced by smaller, cheaper, and more reliable solid-state transistors, the popularity of electronic appliances exploded. In the miniaturized form of the triode transistor, Lee de Forest's invention became the workhorse of our information age.

In the end, de Forest wasn't quite sure whether to be pleased or dismayed by the world he had helped bring into being. In "Dawn of the Electronic Age," a 1952 article he wrote for *Popular Mechanics*, he crowed about his creation of the Audion, referring to it as "this small acorn from which has sprung the gigantic oak that is today world-embracing." At the same time, he lamented the "moral depravity" of commercial broadcast media. "A melancholy view of our national mental level is obtained from a survey of the moronic quality of the majority of today's radio programs," he wrote.

Looking ahead to future applications of electronics, he grew even gloomier. He believed that "electron physiologists" would eventually be able to monitor and analyze "thought or brain waves," allowing "joy and grief [to] be measured in definite, quantitative units." Ultimately, he concluded, "a professor may be able to implant knowledge into the reluctant brains of his 22nd-century pupils. What terrifying political possibilities may be lurking there! Let us be thankful that such things are only for posterity, not for us."[3]

Five

A MEDIUM OF THE MOST
GENERAL NATURE

In the spring of 1954, as the first digital computers were moving into mass production, the brilliant British mathematician Alan Turing killed himself by eating a cyanide-laced apple—a piece of fruit that had been plucked at incalculable cost, the act begs us to conclude, from the tree of knowledge. Turing, who displayed throughout his short life what one biographer calls an "otherworldly innocence,"[1] had during the Second World War played a crucial part in cracking the codes of Enigma, the elaborate typewriter that the Nazis used to encipher and decipher military commands and other sensitive messages. The breaking of Enigma was an epic achievement that helped turn the tide of the war and ensure an Allied victory, though it didn't save Turing from the humiliation of being arrested, a few years later, for having sex with another man.

Today, Alan Turing is best remembered as the creator of an imaginary computing device that anticipated, and served as a blueprint for, the modern computer. He was just twenty-four, a recently elected fellow at Cambridge University, when he introduced what would come to be called the Turing machine in a 1936 paper enti-

tled "On Computable Numbers, with an Application to the Entschei-
dungsproblem." Turing's intent in writing the paper was to show
that there is no such thing as a perfect system of logic or mathematics—that there will always be some statements that cannot be
proven either true or false, that will remain "uncomputable." To help
prove the point, he conjured up a simple, digital calculator able to
follow coded instructions and to read, write, and erase symbols.
Such a computer, he demonstrated, could be programmed to perform the function of any other information-processing device. It was
a "universal machine."[2]

In a later paper, "Computing Machinery and Intelligence," Turing explained how the existence of programmable computers "has
the important consequence that, considerations of speed apart, it is
unnecessary to design various new machines to do various computing processes. They can all be done with one digital computer, suit-
ably programmed for each case." What that means, he concluded, is
that "all digital computers are in a sense equivalent."[3] Turing was not
the first person to imagine how a programmable computer might
work—more than a century earlier, another English mathematician, Charles Babbage, had drawn up plans for an "analytical engine"
that would be "a machine of the most general nature"[4]—but Turing
seems to have been the first to understand the digital computer's
limitless adaptability.

What he could not have anticipated was the way his universal
machine would, just a few decades after his death, become our
universal medium. Because the different sorts of information distributed by traditional media—words, numbers, sounds, images,
moving pictures—can all be translated into digital code, they can all
be "computed." Everything from Beethoven's Ninth to a porn flick
can be reduced to a string of ones and zeros and processed, transmitted, and displayed or played by a computer. Today, with the Internet,
we're seeing firsthand the extraordinary implications of Turing's
discovery. Constructed of millions of interconnected computers and
data banks, the Net is a Turing machine of immeasurable power, and

it is, true to form, subsuming most of our other intellectual technologies. It's becoming our typewriter and our printing press, our map and our clock, our calculator and our telephone, our post office and our library, our radio and our TV. It's even taking over the functions of other computers; more and more of our software programs run through the Internet—or "in the cloud," as the Silicon Valley types say—rather than inside our home computers.

As Turing pointed out, the limiting factor of his universal machine was speed. Even the earliest digital computer could, in theory, do any information-processing job, but a complicated task—rendering a photograph, say—would have taken it far too long, and cost far too much, to be practicable. A guy in a darkroom with trays of chemicals could do the work much more quickly and cheaply. Computing's speed limits, though, turned out to be only temporary obstacles. Since the first mainframe was assembled in the 1940s, the speed of computers and data networks has increased at a breakneck pace, and the cost of processing and transmitting data has fallen equally rapidly. Over the past three decades, the number of instructions a computer chip can process every second has doubled about every three years, while the cost of processing those instructions has fallen by almost half every year. Overall, the price of a typical computing task has dropped by 99.9 percent since the 1960s.[5] Network bandwidth has expanded at an equally fast clip, with Internet traffic doubling, on average, every year since the World Wide Web was invented.[6] Computer applications that were unimaginable in Turing's day are now routine.

The way the Web has progressed as a medium replays, with the velocity of a time-lapse film, the entire history of modern media. Hundreds of years have been compressed into a couple of decades. The first information-processing machine that the Net replicated was Gutenberg's press. Because text is fairly simple to translate into software code and to share over networks—it doesn't require a lot of memory to store, a lot of bandwidth to transmit, or a lot of processing power to render on a screen—early Web sites were usually constructed entirely of typographical symbols. The very term we came to use to describe what we look at online—*pages*—emphasized the connection with printed documents. Publishers of magazines and newspapers, realizing that large quantities of text could, for the first time in history, be broadcast the way radio and TV programs had always been, were among the first businesses to open online outlets, posting articles, excerpts, and other pieces of writing on their sites. The ease with which words could be transmitted led, as well, to the widespread and extraordinarily rapid adoption of e-mail, rendering the personal letter obsolete.

As the cost of memory and bandwidth fell, it became possible to incorporate photographs and drawings into Web pages. At first, the images, like the text they often accompanied, were in black and white, and their low resolution made them blurry. They looked like the first photos printed in newspapers a hundred years ago. But the capacity of the Net expanded to handle color pictures, and the size and quality of the images increased enormously. Soon, simple animations began to play online, mimicking the herky-jerky motions of the flip books, or kineographs, that were popular at the end of the nineteenth century.

Next, the Web began to take over the work of our traditional sound-processing equipment—radios and phonographs and tape decks. The earliest sounds to be heard online were spoken words, but soon snippets of music, and then entire songs and even symphonies, were streaming through sites, at ever-higher levels of fidelity. The network's ability to handle audio streams was aided by the development of software algorithms, such as the one used to produce MP3 files, that erase from music and other recordings sounds that are hard for the human ear to hear. The algorithms allowed sound files to be compressed to much smaller sizes with only slight sacrifices in quality. Telephone calls also began to be routed over the fiber-optic cables of the Internet, bypassing traditional phone lines.

Finally, video came online, as the Net subsumed the technologies of cinema and television. Because the transmission and display of moving pictures place great demands on computers and networks, the first online videos played in tiny windows inside browsers. The pictures would often stutter or drop out, and they were usually out of sync with their soundtracks. But here, too, gains came swiftly. Within just a few years, elaborate three-dimensional games were being played online, and companies like Netflix and Apple were sending high-definition movies and TV shows over the network and onto screens in customers' homes. Even the long-promised "picture phone" is finally becoming a reality, as webcams become a regular feature of computers and Net-connected televisions, and popular Internet telephone services like Skype incorporate video transmissions.

THE NET DIFFERS from most of the mass media it replaces in an obvious and very important way: it's bidirectional. We can send messages through the network as well as receive them. That's made the system all the more useful. The ability to exchange information online, to upload as well as download, has turned the Net into a thoroughfare for business and commerce. With a few clicks, people can search virtual catalogues, place orders, track shipments, and update information in corporate databases. But the Net doesn't just connect us with businesses; it connects us with one another. It's a personal broadcasting medium as well as a commercial one. Millions of people use it to distribute their own digital creations, in the form of blogs, videos, photos, songs, and podcasts, as well as to critique, edit, or otherwise modify the creations of others. The vast, volunteer-written encyclopedia Wikipedia, the largely amateur-produced YouTube video service, the massive Flickr photo repository, the sprawling Huffington Post blog compendium—all of these popular media services were unimaginable before the Web came along. The interactivity of the medium has also turned it into the world's meetinghouse, where people gather to chat, gossip, argue, show off, and flirt on Facebook, Twitter, and all sorts of other social (and sometimes antisocial) networks.

As the uses of the Internet have proliferated, the time we devote to the medium has grown apace, even as speedier connections have allowed us to do more during every minute we're logged on. By 2009, adults in North America were spending an average of twelve hours online a week, double the average in 2005.[7] If you consider only those adults with Internet access, online hours jump considerably, to more than seventeen a week. For younger adults, the figure is higher still, with people in their twenties spending more than nineteen hours a week online.[8] American children between the ages of two and eleven were using the Net about eleven hours a week in 2009, an increase of more than sixty percent since 2004.[9] The typical European adult was online nearly eight hours a week in 2009, up about thirty percent since 2005. Europeans in their twenties were online about twelve hours a week on average.[10] A 2008 international survey of 27,500 adults between the ages of eighteen and fifty-five found that people are spending thirty percent of their leisure time online, with the Chinese being the most intensive surfers, devoting forty-four percent of their off-work hours to the Net.[11]

These figures don't include the time people spend using their mobile phones and other handheld computers to exchange text messages, which also continues to increase rapidly. Text messaging now represents one of the most common uses of computers, particularly for the young. By the beginning of 2009, the average American cell phone user was sending or receiving nearly 400 texts a month, more than a fourfold increase from 2006. The average American teen was sending or receiving a mind-boggling 2,272 texts a month.[12] Worldwide, well over two trillion text messages zip between mobile phones every year, far outstripping the number of voice calls.[13] Thanks to our ever-present messaging systems and devices, we "never really have to disconnect," says Danah Boyd, a social scientist who works for Microsoft.[14]

It's often assumed that the time we devote to the Net comes out of the time we would otherwise spend watching TV. But statistics suggest otherwise. Most studies of media activity indicate that as

Net use has gone up, television viewing has either held steady or increased. The Nielsen Company's long-running media-tracking survey reveals that the time Americans devote to TV viewing has been going up throughout the Web era. The hours we spend in front of the tube rose another two percent between 2008 and 2009, reaching 153 hours a month, the highest level since Nielsen began collecting data in the 1950s (and that doesn't include the time people spend watching TV shows on their computers).[15] In Europe as well, people continue to watch television as much as they ever have. The average European viewed more than a dozen hours of TV a week in 2009, nearly an hour more than in 2004.[16]

A 2006 study by Jupiter Research revealed "a huge overlap" between TV viewing and Web surfing, with forty-two percent of the most avid TV fans (those watching thirty-five or more hours of programming a week) also being among the most intensive users of the Net (those spending thirty or more hours online a week).[17] The growth in our online time has, in other words, expanded the total amount of time we spend in front of screens. According to an extensive 2009 study conducted by Ball State University's Center for Media Design, most Americans, no matter what their age, spend at least eight and a half hours a day looking at a television, a computer monitor, or the screen of their mobile phone. Frequently, they use two or even all three of the devices simultaneously.[18]

What does seem to be decreasing as Net use grows is the time we spend reading print publications—particularly newspapers and magazines, but also books. Of the four major categories of personal media, print is now the least used, lagging well behind television, computers, and radio. By 2008, according to the U.S. Bureau of Labor Statistics, the time that the average American over the age of fourteen devoted to reading printed works had fallen to 143 minutes a week, a drop of eleven percent since 2004. Young adults between the ages of twenty-five and thirty-four, who are among the most avid Net users, were reading printed works for a total of just forty-nine minutes a week in 2008, down a precipitous twenty-nine percent from 2004.[19] In a small but telling 2008 study conducted for *Adweek* magazine, four typical Americans—a barber, a chemist, an elementary school principal, and a real estate agent—were shadowed during the course of a day to document their media usage. The people displayed very different habits, but they shared one thing in common, according to the magazine: "None of the four cracked open any print media during their observed hours."[20] Because of the ubiquity of text on the Net and our phones, we're almost certainly reading more words today than we did twenty years ago, but we're devoting much less time to reading words printed on paper.

The Internet, like the personal computer before it, has proven to be so useful in so many ways that we've welcomed every expansion of its scope. Rarely have we paused to ponder, much less question, the media revolution that has been playing out all around us, in our homes, our workplaces, our schools. Until the Net arrived, the history of media had been a tale of fragmentation. Different technologies progressed down different paths, leading to a proliferation of special-purpose tools. Books and newspapers could present text and images, but they couldn't handle sounds or moving pictures. Visual media like cinema and TV were unsuited to the display of text, except in the smallest of quantities. Radios, telephones, phonographs, and tape players were limited to transmitting sounds. If you wanted to add up numbers, you used a calculator. If you wanted to look up facts, you consulted a set of encyclopedias or a *World Almanac*. The production end of the business was every bit as fragmented as the consumption end. If a company wanted to sell words, it printed them on paper. If it wanted to sell movies, it wound them onto spools of film. If it wanted to sell songs, it pressed them onto vinyl records or recorded them onto magnetic tape. If it wanted to distribute TV shows and commercials, it shot them through the air from a big antenna or sent them down thick black coaxial cables.

Once information is digitized, the boundaries between media dissolve. We replace our special-purpose tools with an all-purpose tool. And because the economics of digital production and distribution are almost always superior to what came before—the cost of creating electronic products and transmitting them through the Net is a small fraction of the cost of manufacturing physical goods and shipping them through warehouses and into stores—the shift happens very quickly, following capitalism's inexorable logic. Today, nearly all media companies distribute digital versions of their products through the Net, and the growth in the consumption of media goods is taking place almost entirely online.

That doesn't mean that traditional forms of media have disappeared. We still buy books and subscribe to magazines. We still go to the movies and listen to the radio. Some of us still buy music on CDs and movies on DVDs. A few of us will even pick up a newspaper now and then. When old technologies are supplanted by new ones, the old technologies often continue to be used for a long time, sometimes indefinitely. Decades after the invention of movable type, many books were still being handwritten by scribes or printed from woodblocks—and some of the most beautiful books continue to be produced in those ways today. Quite a few people still listen to vinyl records, use film cameras to take photographs, and look up phone numbers in the printed Yellow Pages. But the old technologies lose their economic and cultural force. They become progress's dead ends. It's the new technologies that govern production and consumption, that guide people's behavior and shape their perceptions. That's why the future of knowledge and culture no longer lies in books or newspapers or TV shows or radio programs or records or CDs. It lies in digital files shot through our universal medium at the speed of light.

"A NEW MEDIUM is never an addition to an old one," wrote McLuhan in *Understanding Media*, "nor does it leave the old one in peace. It never ceases to oppress the older media until it finds new shapes and positions for them."[21] His observation rings particularly true today. Traditional media, even electronic ones, are being refashioned and repositioned as they go through the shift to online distribution.

When the Net absorbs a medium, it re-creates that medium in its own image. It not only dissolves the medium's physical form; it injects the medium's content with hyperlinks, breaks up the content into searchable chunks, and surrounds the content with the content of all the other media it has absorbed. All these changes in the form of the content also change the way we use, experience, and even understand the content.

A page of online text viewed through a computer screen may seem similar to a page of printed text. But scrolling or clicking through a Web document involves physical actions and sensory stimuli very different from those involved in holding and turning the pages of a book or a magazine. Research has shown that the cognitive act of reading draws not just on our sense of sight but also on our sense of touch. It's tactile as well as visual. "All reading," writes Anne Mangen, a Norwegian literary studies professor, is "multi-sensory." There's "a crucial link" between "the sensory-motor experience of the materiality" of a written work and "the cognitive processing of the text content."[22] The shift from paper to screen doesn't just change the way we navigate a piece of writing. It also influences the degree of attention we devote to it and the depth of our immersion in it.

Hyperlinks also alter our experience of media. Links are in one sense a variation on the textual allusions, citations, and footnotes that have long been common elements of documents. But their effect on us as we read is not at all the same. Links don't just point us to related or supplemental works; they propel us toward them. They encourage us to dip in and out of a series of texts rather than devote sustained attention to any one of them. Hyperlinks are designed to grab our attention. Their value as navigational tools is inextricable from the distraction they cause.

The searchability of online works also represents a variation on older navigational aids such as tables of contents, indexes, and concordances. But here, too, the effects are different. As with links, the ease and ready availability of searching make it much simpler to jump between digital documents than it ever was to jump between printed ones. Our attachment to any one text becomes more tenuous, more provisional. Searches also lead to the fragmentation of online works. A search engine often draws our attention to a particular snippet of text, a few words or sentences that have strong relevance to whatever we're searching for at the moment, while providing little incentive for taking in the work as a whole. We don't see the forest when we search the Web. We don't even see the trees. We see twigs and leaves. As companies like Google and Microsoft perfect search engines for video and audio content, more products are undergoing the fragmentation that already characterizes written works.

By combining many different kinds of information on a single screen, the multimedia Net further fragments content and disrupts our concentration. A single Web page may contain a few chunks of text, a video or audio stream, a set of navigational tools, various advertisements, and several small software applications, or "widgets," running in their own windows. We all know how distracting this cacophony of stimuli can be. We joke about it all the time. A new e-mail message announces its arrival as we're glancing over the latest headlines at a newspaper's site. A few seconds later, our RSS reader tells us that one of our favorite bloggers has uploaded a new post. A moment after that, our mobile phone plays the ringtone that signals an incoming text message. Simultaneously, a Facebook or Twitter alert blinks on-screen. In addition to everything flowing through the network, we also have immediate access to all the other software programs running on our computers—they, too, compete for a piece of our mind. Whenever we turn on our computer, we are plunged into an "ecosystem of interruption technologies," as the blogger and science fiction writer Cory Doctorow terms it.[23]

Interactivity, hyperlinking, searchability, multimedia—all these qualities of the Net bring attractive benefits. Along with the unprecedented volume of information available online, they're the main reasons that most of us are drawn to using the Net so much. We like to be able to switch between reading and listening and watching without having to get up and turn on another appliance or dig through a pile of magazines or disks. We like to be able to find and be transported instantly to relevant data—without having to sort through lots of extraneous stuff. We like to be in touch with friends, family members, and colleagues. We like to feel connected—and we hate to feel disconnected. The Internet doesn't change our intellectual habits against our will. But change them it does.

Our use of the Net will only grow, and its impact on us will only strengthen, as it becomes ever more present in our lives. Like the clock and the book before it, the computer continues to get smaller and cheaper as technology advances. Inexpensive laptops gave us the ability to take the Internet with us when we left our office or our home. But the laptop was itself a cumbersome device, and connecting one to the Internet was not always easy. The introduction of the tiny netbook and the even tinier smartphone solves those problems. Powerful pocket-sized computers like the Apple iPhone, the Motorola Droid, and the Google Nexus One come bundled with Internet access. Along with the incorporation of Internet services into everything from car dashboards to televisions to the cabins of airplanes, these small devices promise to more deeply integrate the Web into our everyday activities, making our universal medium all the more universal.

As the Net expands, other media contract. By changing the economics of production and distribution, the Net has cut into the profitability of many news, information, and entertainment businesses, particularly those that have traditionally sold physical products. Sales of music CDs have fallen steadily over the last decade, dropping twenty percent in 2008 alone.[24] Sales of movie DVDs, a major recent source of profits for Hollywood studios, are also now in decline, falling six percent during 2008 and then plunging another fourteen percent during the first half of 2009.[25] Unit sales of greeting cards and postcards are dropping.[26] The volume of mail sent through the U.S. Postal Service declined at its fastest pace ever during 2009.[27] Universities are discontinuing the printed editions of scholarly monographs and journals and moving to strictly electronic distribution.[28] Public schools are pushing students to use online reference materials in place of what California Governor Arnold Schwarzenegger refers to as "antiquated, heavy, expensive textbooks."[29] Everywhere you look, you see signs of the Net's growing hegemony over the packaging and flow of information.

Nowhere have the effects been so unsettling as in the newspaper industry, which faces particularly severe financial challenges as readers and advertisers embrace the Net as their medium of choice. The decline in Americans' newspaper reading began decades ago, when radio and TV began consuming more of peoples' leisure time, but the Internet has accelerated the trend. Between 2008 and 2009, newspaper circulation dropped more than seven percent, while visits to newspaper Web sites grew by more than ten percent.[30] One of America's oldest dailies, the *Christian Science Monitor*, announced in early 2009 that after a hundred years it was stopping its presses. The Web would become its main channel for distributing news. The move, said the paper's publisher, Jonathan Wells, was a harbinger of what lay in store for other newspapers. "Changes in the industry—changes in the concept of news and the economics underlying the industry—hit the *Monitor* first," he explained.[31]

He was soon proved correct. Within months, Colorado's oldest newspaper, the *Rocky Mountain News*, had gone out of business; the *Seattle Post-Intelligencer* had abandoned its print edition and fired most of its staff; the *Washington Post* had shut down all its U.S. bureaus and let more than a hundred journalists go; and the owners of more than thirty other U.S. newspapers, including the *Los Angeles Times*, *Chicago Tribune*, *Philadelphia Inquirer*, and *Minneapolis Star Tribune*, had filed for bankruptcy. Tim Brooks, the managing director of Guardian News and Media, which publishes *The Guardian* and *The Independent* in Britain, announced that all his company's future investments would go into multimedia digital products, mainly delivered through its Web sites. "The days when you can trade in just words are gone," he told an industry conference.[32]

AS PEOPLE'S MINDS become attuned to the crazy quilt of Web content, media companies have to adapt to the audience's new expectations. Many producers are chopping up their products to fit the shorter attention spans of online consumers, as well as to raise their profiles on search engines. Snippets of TV shows and movies are distributed through YouTube, Hulu, and other video services. Excerpts of radio programs are offered as podcasts or streams. Individual magazine and newspaper articles circulate in isolation. Pages of books are displayed through Amazon.com and Google Book Search. Music albums are split apart, their songs sold through iTunes or streamed through Spotify. Even the songs themselves are broken into pieces, with their riffs and hooks packaged as ringtones for cell phones or embedded in video games. There's much to be said for what economists call the "unbundling" of content. It provides people with more choices and frees them from unwanted purchases. But it also illustrates and reinforces the changing patterns of media consumption promoted by the Web. As the economist Tyler Cowen says, "When access [to information] is easy, we tend to favor the short, the sweet, and the bitty."[33]

The Net's influence doesn't end at the edge of a computer screen. Media companies are reshaping their traditional products, even the physical ones, to more closely resemble what people experience when they're online. If, in the early days of the Web, the design of online publications was inspired by print publications (as the design of Gutenberg's Bible was inspired by scribal books), today the inspiration tends to go in the opposite direction. Many magazines have tweaked their layouts to mimic or at least echo the look and feel of Web sites. They've shortened their articles, introduced capsule summaries, and crowded their pages with easy-to-browse blurbs and captions. *Rolling Stone*, once known for publishing sprawling, adventurous features by writers like Hunter S. Thompson, now eschews such works, offering readers a jumble of short articles and reviews. There was "no Internet," publisher Jann Wenner explains, "back when *Rolling Stone* was publishing these seven-thousand-word stories." Most popular magazines have come to be "filled with color, oversized headlines, graphics, photos, and pull quotes," writes Michael Scherer in the *Columbia Journalism Review*. "The gray text page, once a magazine staple, has been all but banished."[34]

The design of newspapers is also changing. Many papers, including industry stalwarts like the *Wall Street Journal* and the *Los Angeles Times*, have over the last few years moved to trim the length of their articles and introduce more summaries and navigational aids to make the scanning of their contents easier. An editor at the *Times* of London attributes such format changes to the newspaper industry's adaptation to "an Internet age, a headline age."[35] In March of 2008, the *New York Times* announced it would begin devoting three pages of every edition to paragraph-long article abstracts and other brief items. Its design director, Tom Bodkin, explained that the "shortcuts" would allow harried readers to get a quick "taste" of the day's news, sparing them the "less efficient" method of actually turning the pages and reading the articles.[36]

Such copycat strategies haven't been particularly successful in stanching the flow of readers from print to online publications. After a year, during which its circulation continued to decline, the *New York Times* quietly abandoned much of its redesign, restricting article summaries to a single page in most editions. A few magazines, realizing that competing with the Web on its own terms is a losing proposition, have reversed their strategies. They've gone back to simpler, less cluttered designs and longer articles. *Newsweek* overhauled its pages in 2009, placing a greater emphasis on essays and professional photographs and adopting a heavier, more expensive paper stock. The price that publications pay for going against the conventions of the Web is a further whittling of their readership. When *Newsweek* unveiled its new design, it also announced it was slashing the circulation it guaranteed its advertisers from 2.6 million to 1.5 million.[37]

Like their print counterparts, most TV shows and movies are also trying to become more Web-like. Television networks have added text "crawls" and "flippers" to their screens and routinely run info-graphics and pop-up ads during their programs. Some newer shows, such as NBC's *Late Night with Jimmy Fallon*, have been explicitly designed to cater as much to Net surfers as TV viewers, with an emphasis on brief segments that lend themselves to distribution as YouTube clips. Cable and satellite companies offer theme channels that enable viewers to watch several programs simultaneously, using their remote control as a kind of mouse to click between audio tracks. Web content is also beginning to be offered directly through TVs, as leading television manufacturers like Sony and Samsung redesign their sets to seamlessly combine Internet programming with traditional broadcasts. Movie studios have begun incorporating social-networking features into the disks they sell. With the Blu-ray version of Disney's *Snow White*, viewers can chat with one another through the Net while watching the seven dwarves march off to work. The disk of *Watchmen* automatically syncs with Facebook accounts, letting viewers exchange "live commentary" on the film with their "friends."[38] Craig Kornblau, the president of Universal Studios Home Entertainment, says the studio plans to introduce more such features, with the goal of turning the viewing of movies into "interactive experiences."[39]

The Net has begun to alter the way we experience actual performances as well as the recordings of those performances. When we carry a powerful mobile computer into a theater or other venue, we carry, as well, all the communication and social-networking tools available on the Web. It long ago became common for concertgoers to record and broadcast snippets of shows to friends through the cameras in their cell phones. Now, mobile computers are beginning to be deliberately incorporated into performances as a way to appeal to a new generation of Net-saturated patrons. During a 2009 performance of Beethoven's *Pastoral* Symphony at Wolf Trap in Virginia, the National Symphony Orchestra sent out a stream of Twitter tweets, written by conductor Emil de Cou, explaining some of Beethoven's musical references.[40] The New York Philharmonic and the Indianapolis Symphony Orchestra have begun encouraging audience members to use their phones

to vote, via text messaging, for the evening's encore. "It was less passive than just sitting there and listening to music," commented an
attendee after a recent Philharmonic performance.[41] A growing number of American churches are encouraging parishioners to bring laptops and smartphones to services in order to exchange inspirational
messages through Twitter and other microblogging services.[42] Eric
Schmidt, Google's chief executive, sees the incorporation of social
networking into theatrical and other events as an exciting new business opportunity for Internet firms. "The most obvious use of Twitter," he says, can be seen in situations where "everybody is watching a
play and are busy talking about the play while the play is under way."[43]
Even the experiences we have in the real world are coming to be mediated by networked computers.

A particularly striking illustration of how the Net is reshaping our
expectations about media can be seen in any library. Although we
don't tend to think of libraries as media technologies, they are. The
public library is, in fact, one of the most important and influential
informational media ever created—and one that proliferated only
after the arrival of silent reading and movable-type printing. A community's attitudes and preferences toward information take concrete
shape in its library's design and services. Until recently, the public
library was an oasis of bookish tranquility where people searched
through shelves of neatly arranged volumes or sat in carrels and read
quietly. Today's library is very different. Internet access is rapidly
becoming its most popular service. According to recent surveys by
the American Library Association, ninety-nine percent of U.S. public
library branches provide Internet access, and the average branch has
eleven public computers. More than three-quarters of branches also
offer Wi-Fi networks for their patrons' use.[44] The predominant sound
in the modern library is the tapping of keys, not the turning of pages.

The architecture of one of the newest branches of the venerable
New York Public Library, the Bronx Library Center, testifies to the
library's changing role. Writing in the journal *Strategy & Business*,
three management consultants describe the building's layout: "On

the library's four main floors, the stacks of books have been placed at each end, leaving ample space in the middle for tables that have computers on them, many with broadband access to the Internet. The people using the computers are young and aren't necessarily using them for academic purposes—here is one doing a Google search on Hannah Montana pictures, there is one updating his Facebook page, and over there a few children are playing video games, including The Fight for Glorton. Librarians answer questions and organize online gaming tournaments, and none of them are shushing anyone."[45] The consultants point to the Bronx branch as an example of how forward-looking libraries are retaining their "relevance" by "launching new digital initiatives to meet users' needs." The library's layout provides, as well, a powerful symbol of our new media landscape: at the center stands the screen of the Internet-connected computer; the printed word has been pushed to the margins.

Six

THE VERY IMAGE OF A BOOK

And what of the book itself? Of all popular media, it's probably the one that has been most resistant to the Net's influence. Book publishers have suffered some losses of business as reading has shifted from the printed page to the screen, but the form of the book itself hasn't changed much. A long sequence of printed pages assembled between a pair of stiff covers has proven to be a remarkably robust technology, remaining useful and popular for more than half a millennium.

It's not hard to see why books have been slow to make the leap into the digital age. There's not a whole lot of difference between a computer monitor and a television screen, and the sounds coming from speakers hit your ears in pretty much the same way whether they're being transmitted through a computer or a radio. But as a device for reading, the book retains some compelling advantages over the computer. You can take a book to the beach without worrying about sand getting in its works. You can take it to bed without being nervous about it falling to the floor should you nod off. You can spill coffee on it. You can sit on it. You can put it down on a table, open to the page you're reading, and when you pick it up a few days later it will still be exactly as you left it. You never have to be concerned about plugging a book into an outlet or having its battery die.

The experience of reading tends to be better with a book too. Words stamped on a page in black ink are easier to read than words formed of pixels on a backlit screen. You can read a dozen or a hundred printed pages without suffering the eye fatigue that often results from even a brief stretch of online reading. Navigating a book is simpler and, as software programmers say, more intuitive. You can flip through real pages much more quickly and flexibly than you can through virtual pages. And you can write notes in a book's margins or highlight passages that move or inspire you. You can even get a book's author to sign its title page. When you're finished with a book, you can use it to fill an empty space on your bookshelf—or lend it to a friend.

Despite years of hype about electronic books, most people haven't shown much interest in them. Investing a few hundred dollars in a specialized "digital reader" has seemed silly, given the ease and pleasure of buying and reading old-fashioned books. But books will not remain exempt from the digital media revolution. The economic advantages of digital production and distribution—no big purchases of ink and paper, no printer bills, no loading of heavy boxes onto trucks, no returns of unsold copies—are every bit as compelling for book publishers and distributors as for other media companies. And the lower costs translate into lower prices. It's not unusual for e-books to be sold for half the price of print editions, thanks in part to subsidies from device manufacturers. The sharp discounts provide a strong incentive for people to make the switch from paper to pixels.

Digital readers have also improved greatly in recent years. The advantages of traditional books are not quite as clear-cut as they used to be. Thanks to high-resolution screens made of materials like Vizplex, a charged-particle film developed by the Massachusetts company E Ink, the clarity of digital text now almost rivals that of printed text. The latest readers don't require backlighting, allowing them to be used in direct sunlight and reducing eye strain considerably. The functions of the readers have also improved, making it much easier to click through pages, add bookmarks, highlight text, and even scribble marginal notes. People with weak eyes can increase the size of the type in e-books—something they can't do with printed books. And as computer memory prices have gone down, the capacity of the readers has gone up. You can now load them with hundreds of books. Just as an iPod can hold the entire contents of an average person's music collection, so an e-book reader can now hold an entire personal library.

Although sales of e-books still represent a tiny fraction of overall book sales, they have been increasing at a much faster pace than sales of physical books. Amazon.com reported in early 2009 that for the 275,000 books it sells in both traditional and digital form, the e-book versions account for thirty-five percent of total sales, up sharply from less than ten percent just a year earlier. Long stagnant, sales of digital readers are now booming, rising from about one million units in 2008 to an estimated twelve million in 2010.[1] As Brad Stone and Motoko Rich of the *New York Times* recently reported, "the e-book has started to take hold."[2]

ONE OF THE more popular of the new digital readers is Amazon's own Kindle. Introduced with great fanfare in 2007, the gadget incorporates all the latest screen technology and reading functions and includes a full keypad. But it has another feature that greatly increases its attractiveness. The Kindle has a built-in, always-available wireless connection to the Internet. The cost of the connection is rolled into the price of the Kindle, so there's no additional subscription fee involved. The connection allows you, not surprisingly, to shop for books at the Amazon store and immediately download the ones you buy. But it lets you do much more than that. You can read digital newspapers and magazines, scan blogs, perform Google searches, listen to MP3s, and, through a specially made browser, surf other Web sites. The Kindle's most radical feature, at least when it comes to thinking about what's in store for books, is its incorporation of links into the text it displays. The Kindle turns the words of books into hypertext. You can click on a word or a phrase and be taken to a related dictionary entry, Wikipedia article, or list of Google search results.

The Kindle points to the future of digital readers. Its features, and even its software, are being incorporated into iPhones and PCs, transforming the reader from a specialized and expensive device to just another cheap application running in Turing's universal machine. The Kindle also, if less happily, points to the future of books. In a 2009 *Newsweek* article, the journalist and editor Jacob Weisberg, once a skeptic about electronic books, praised the Kindle as "a machine that marks a cultural revolution" in which "reading and printing are getting separated." What the Kindle tells us, Weisberg went on, is "that printed books, *the most important artifacts of human civilization*, are going to join newspapers and magazines on the road to obsolescence."[3] Charles McGrath, onetime editor of the *New York Times Book Review*, has also become a Kindle believer, calling "the seductive white gizmo" a "precursor" of what's to come for books and reading. "It's surprising how easily you succumb to convenience," he says, "and how little you miss, once they're gone, all the niceties of typography and design that you used to value so much." While he doesn't think that printed books are going to disappear anytime soon, he does sense that "in the future we will keep them around as fond relics, reminders of what reading used to be like."[4]

What would that mean for how we read what we used to read in books? The *Wall Street Journal*'s L. Gordon Crovitz has suggested that easy-to-use, networked readers like the Kindle "can help return to us our attention spans and extend what makes books great: words and their meaning."[5] That's a sentiment most literary-minded folks would be eager to share. But it's wishful thinking. Crovitz has fallen victim to the blindness that McLuhan warned against: the inability to see how a change in a medium's form is also a change in its content. "E-books should not just be print books delivered electronically," says a senior vice president of HarperStudio, an imprint of the publishing giant HarperCollins. "We need to take advantage of the medium and create something dynamic to enhance the experience. I want links and behind the scenes extras and narration and videos and conversation."[6] As soon as you inject a book with links and connect it to the Web—as soon as you "extend" and "enhance" it and make it "dynamic"—you change what it is and you change, as well, the experience of reading it. An e-book is no more a book than an online newspaper is a newspaper.

Soon after the author Steven Johnson began reading e-books on his new Kindle, he realized that "the book's migration to the digital realm would not be a simple matter of trading ink for pixels, but would likely change the way we read, write, and sell books in profound ways." He was excited by the Kindle's potential for expanding "the universe of books at our fingertips" and making books as searchable as Web pages. But the digital device also filled him with trepidation: "I fear that one of the great joys of book reading—the total immersion in another world, or in the world of the author's ideas—will be compromised. We all may read books the way we increasingly read magazines and newspapers: a little bit here, a little bit there."[7]

Christine Rosen, a fellow at the Ethics and Public Policy Center in Washington, DC, recently wrote about her experience using a Kindle to read the Dickens novel *Nicholas Nickleby*. Her story underscores Johnson's fears: "Although mildly disorienting at first, I quickly adjusted to the Kindle's screen and mastered the scroll and page-turn buttons. Nevertheless, my eyes were restless and jumped around as they do when I try to read for a sustained time on the computer. Distractions abounded. I looked up Dickens on Wikipedia, then jumped straight down the Internet rabbit hole following a link about a Dickens short story, 'Mugby Junction.' Twenty minutes later I still hadn't returned to my reading of *Nickleby* on the Kindle."[8]

Rosen's struggle sounds almost identical to the one that the historian David Bell went through back in 2005 when he read a new electronic book, *The Genesis of Napoleonic Propaganda*, on the Internet. He described his experience in a *New Republic* article: "A few clicks, and the text duly appears on my computer screen. I start reading, but while the book is well written and informative, I find it remarkably hard to concentrate. I scroll back and forth, search for key words, and interrupt myself even more often than usual to refill my coffee cup, check my e-mail, check the news, rearrange files in my desk drawer. Eventually I get through the book and am glad to have done so. But a week later I find it remarkably hard to remember what I have read."[9]

When a printed book—whether a recently published scholarly history or a two-hundred-year-old Victorian novel—is transferred to an electronic device connected to the Internet, it turns into something very like a Web site. Its words become wrapped in all the distractions of the networked computer. Its links and other digital enhancements propel the reader hither and yon. It loses what the late John Updike called its "edges" and dissolves into the vast, roiling waters of the Net.[10] The linearity of the printed book is shattered, along with the calm attentiveness it encourages in the reader. The high-tech features of devices like the Kindle and Apple's new iPad may make it more likely that we'll read e-books, but the way we read them will be very different from the way we read printed editions.

CHANGES IN READING style will also bring changes in writing style, as authors and their publishers adapt to readers' new habits and expectations. A striking example of this process is already on display in Japan. In 2001, young Japanese women began composing stories on their mobile phones, as strings of text messages, and uploading them to a Web site, Maho no i-rando, where other people read and commented on them. The stories expanded into serialized "cell phone novels," and their popularity grew. Some of the novels found millions of readers online. Publishers took notice, and began to bring out the novels as printed books. By the end of the decade, cell phone novels had come to dominate the country's best-seller lists. The three top-selling Japanese novels in 2007 were all originally written on mobile phones.

The form of the novels reflects their origins. They are, according to the reporter Norimitsu Onishi, "mostly love stories written in the short sentences characteristic of text messaging but containing little of the plotting or character development found in traditional novels." One of the most popular cell phone novelists, a twenty-one-year-old who goes by the name of Rin, explained to Onishi why young readers are abandoning traditional novels: "They don't read works by professional writers because their sentences are too difficult to understand, their expressions are intentionally wordy, and the stories are not familiar to them."[11] The popularity of cell phone novels may never extend beyond Japan, a country given to peculiar fads, but the novels nevertheless demonstrate how changes in reading inevitably spur changes in writing.

Another sign of how the Web is beginning to influence book writing came in 2009, when O'Reilly Media, an American publisher of technology books, brought out a book about Twitter that had been created with Microsoft's PowerPoint presentation software. "We've long been interested in exploring how the online medium changes the presentation, narrative and structure of the book," said the firm's chief executive, Tim O'Reilly, in introducing the volume, which is available in both print and electronic editions. "Most books still use the old model of a sustained narrative as their organizational principle. Here, we've used a web-like model of standalone pages, each of which can be read alone (or at most in a group of two or three)." The "modular architecture" reflects the way people's reading practices have changed as they've adapted to online text, O'Reilly explained. The Web "provides countless lessons about how books need to change when they move online."[12]

Some of the changes in the way books are written and presented will be dramatic. At least one major publisher, Simon & Schuster, has already begun publishing e-novels that have videos embedded in their virtual pages. The hybrids are known as "vooks." Other companies have similar multimedia experiments in the works. "Everybody is trying to think about how books and information will best be put together in the 21st century," said Simon & Schuster executive Judith

Curr in explaining the impetus behind vooks. "You can't just be linear anymore with your text."[13]

Other changes in form and content will be subtle, and they'll develop slowly. As more readers come to discover books through online text searches, for example, authors will face growing pressures to tailor their words to search engines, the way bloggers and other Web writers routinely do today. Steven Johnson sketches out some of the likely consequences: "Writers and publishers will begin to think about how individual pages or chapters might rank in Google's results, crafting sections explicitly in the hopes that they will draw in that steady stream of search visitors. Individual paragraphs will be accompanied by descriptive tags to orient potential searchers; chapter titles will be tested to determine how well they rank."[14]

Many observers believe it's only a matter of time before social-networking functions are incorporated into digital readers, turning reading into something like a team sport. We'll chat and pass virtual notes while scanning electronic text. We'll subscribe to services that automatically update our e-books with comments and revisions added by fellow readers. "Soon," says Ben Vershbow of the Institute for the Future of the Book, an arm of USC's Annenberg Center for Communication, "books will literally have discussions inside of them, both live chats and asynchronous exchanges through comments and social annotation. You will be able to see who else out there is reading that book and be able to open up a dialog with them."[15] In a much-discussed essay, the science writer Kevin Kelly even suggested that we'll be holding communal cut-and-paste parties online. We'll cobble together new books from bits and pieces lifted out of old ones. "Once digitized," he wrote, "books can be unraveled into single pages or be reduced further, into snippets of a page. These snippets will be remixed into reordered books," which will then "be published and swapped in the public commons."[16]

That particular scenario may or may not come to pass, but it does seem inevitable that the Web's tendency to turn all media into social media will have a far-reaching effect on styles of reading and writing and hence on language itself. When the form of the book shifted to accommodate silent reading, one of the most important results was the development of private writing. Authors, able to assume that an attentive reader, deeply engaged both intellectually and emotionally, "would come at last, and would thank them," quickly jumped beyond the limits of social speech and began to explore a wealth of distinctively literary forms, many of which could exist only on the page. The new freedom of the private writer led, as we've seen, to a burst of experimentation that expanded vocabulary, extended the boundaries of syntax, and in general increased the flexibility and expressiveness of language. Now that the context of reading is again shifting, from the private page to the communal screen, authors will adapt once more. They will increasingly tailor their work to a milieu that the essayist Caleb Crain describes as "groupiness," where people read mainly "for the sake of a feeling of belonging" rather than for personal enlightenment or amusement.[17] As social concerns override literary ones, writers seem fated to eschew virtuosity and experimentation in favor of a bland but immediately accessible style. Writing will become a means for recording chatter.

The provisional nature of digital text also promises to influence writing styles. A printed book is a finished object. Once inked onto the page, its words become indelible. The finality of the act of publishing has long instilled in the best and most conscientious writers and editors a desire, even an anxiety, to perfect the works they produce—to write with an eye and an ear toward eternity. Electronic text is impermanent. In the digital marketplace, publication becomes an ongoing process rather than a discrete event, and revision can go on indefinitely. Even after an e-book is downloaded into a networked device, it can be easily and automatically updated—just as software programs routinely are today.[18] It seems likely that removing the sense of closure from book writing will, in time, alter writers' attitudes toward their work. The pressure to achieve perfection will diminish, along with the artistic rigor that the pressure imposed. To see how small changes in writers' assumptions and attitudes can eventually have large effects on what they write, one need only glance at the history of correspondence. A personal letter written in, say, the nineteenth century bears little resemblance to a personal e-mail or text message written today. Our indulgence in the pleasures of informality and immediacy has led to a narrowing of expressiveness and a loss of eloquence.[19]

No doubt the connectivity and other features of e-books will bring new delights and diversions. We may even, as Kelly suggests, come to see digitization as a liberating act, a way of freeing text from the page. But the cost will be a further weakening, if not a final severing, of the intimate intellectual attachment between the lone writer and the lone reader. The practice of deep reading that became popular in the wake of Gutenberg's invention, in which "the quiet was part of the meaning, part of the mind," will continue to fade, in all likelihood becoming the province of a small and dwindling elite. We will, in other words, revert to the historical norm. As a group of Northwestern University professors wrote in a 2005 article in the *Annual Review of Sociology*, the recent changes in our reading habits suggest that the "era of mass [book] reading" was a brief "anomaly" in our intellectual history: "We are now seeing such reading return to its former social base: a self-perpetuating minority that we shall call the reading class." The question that remains to be answered, they went on, is whether that reading class will have the "power and prestige associated with an increasingly rare form of cultural capital" or will be viewed as the eccentric practitioners of "an increasingly arcane hobby."[20]

When Amazon's chief executive, Jeff Bezos, introduced the Kindle, he sounded a self-congratulatory note: "It's so ambitious to take something as highly evolved as a book and improve on it. And maybe even change the way people read."[21] There's no "maybe" about it. The way people read—and write—has already been changed by the Net, and the changes will continue as, slowly but surely, the words of books are extracted from the printed page and embedded in the computer's "ecosystem of interruption technologies."

PUNDITS HAVE BEEN trying to bury the book for a long time. In the early years of the nineteenth century, the burgeoning popularity of newspapers—well over a hundred were being published in London alone—led many observers to assume that books were on the verge of obsolescence. How could they compete with the immediacy of the daily broadsheet? "Before this century shall end, journalism will be the whole press—the whole human thought," declared the French poet and politician Alphonse de Lamartine in 1831. "Thought will spread across the world with the rapidity of light, instantly conceived, instantly written, instantly understood. It will blanket the earth from one pole to the other—sudden, instantaneous, burning with the fervor of the soul from which it burst forth. This will be the reign of the human word in all its plenitude. Thought will not have time to ripen, to accumulate into the form of a book—the book will arrive too late. The only book possible from today is a newspaper."[22]

Lamartine was mistaken. At the century's end, books were still around, living happily beside newspapers. But a new threat to their existence had already emerged: Thomas Edison's phonograph. It seemed obvious, at least to the intelligentsia, that people would soon be listening to literature rather than reading it. In an 1889 essay in the *Atlantic Monthly*, Philip Hubert predicted that "many books and stories may not see the light of print at all; they will go into the hands of their readers, or hearers rather, as phonograms." The phonograph, which at the time could record sounds as well as play them, also "promises to far outstrip the typewriter" as a tool for composing prose, he wrote.[23] That same year, the futurist Edward Bellamy suggested, in a *Harper's* article, that people would come to read "with the eyes shut." They would carry around a tiny audio player, called an "indispensable," which would contain all their books, newspapers, and magazines. Mothers, wrote Bellamy, would no longer have "to make themselves hoarse telling the children stories on rainy days to keep them out of mischief." The kids would all have their own indispensables.[24]

Five years later, *Scribner's Magazine* delivered the seeming coup de grâce to the codex, publishing an article titled "The End of Books" by Octave Uzanne, an eminent French author and publisher. "What is my view of the destiny of books, my dear friends?" he wrote. "I do not believe (and the progress of electricity and modern mechanism forbids me to believe) that Gutenberg's invention can do otherwise than sooner or later fall into desuetude as a means of current interpretation of our mental products." Printing, a "somewhat antiquated process" that for centuries "has reigned despotically over the mind of man," would be replaced by "phonography," and libraries would be turned into "phonographotecks." We would see a return of "the art of utterance," as narrators took the place of writers. "The ladies," Uzanne concluded, "will no longer say in speaking of a successful author, 'What a charming writer!' All shuddering with emotion, they will sigh, 'Ah, how this "Teller's" voice thrills you, charms you, moves you.'"[25]

The book survived the phonograph as it had the newspaper. Listening didn't replace reading. Edison's invention came to be used mainly for playing music rather than declaiming poetry and prose. During the twentieth century, book reading would withstand a fresh onslaught of seemingly mortal threats: moviegoing, radio listening, TV viewing. Today, books remain as commonplace as ever, and there's every reason to believe that printed works will continue to be produced and read, in some sizable quantity, for years to come. While physical books may be on the road to obsolescence, the road will almost certainly be a long and winding one. Yet the continued existence of the codex, though it may provide some cheer to bibliophiles, doesn't change the fact that books and book reading, at least as we've defined those things in the past, are in their cultural twilight. As a society, we devote ever less time to reading printed words, and even when we do read them, we do so in the busy shadow of the Internet. "Already," the literary critic George Steiner wrote in 1997, "the silences, the arts of concentration and memorization, the luxuries of time on which 'high reading' depended are largely disposed." But "these erosions," he continued, "are nearly insignificant compared with the brave new world of the electronic."[26] Fifty years ago, it would have been possible to make the case that we were still in the age of print. Today, it is not.

Some thinkers welcome the eclipse of the book and the literary mind it fostered. In a recent address to a group of teachers, Mark Federman, an education researcher at the University of Toronto, argued that literacy, as we've traditionally understood it, "is now nothing but a quaint notion, an aesthetic form that is as irrelevant to the real questions and issues of pedagogy today as is recited poetry—clearly not devoid of value, but equally no longer the structuring force of society." The time has come, he said, for teachers and students alike to abandon the "linear, hierarchical" world of the book and enter the Web's "world of ubiquitous connectivity and pervasive proximity"—a world in which "the greatest skill" involves "discovering emergent meaning among contexts that are continually in flux."[27]

Clay Shirky, a digital-media scholar at New York University, suggested in a 2008 blog post that we shouldn't waste our time mourning the death of deep reading—it was overrated all along. "No one reads *War and Peace*," he wrote, singling out Tolstoy's epic as the quintessence of high literary achievement. "It's too long, and not so interesting." People have "increasingly decided that Tolstoy's sacred work isn't actually worth the time it takes to read it." The same goes for Proust's *In Search of Lost Time* and other novels that until recently were considered, in Shirky's cutting phrase, "Very Important in some vague way." Indeed, we've "been emptily praising" writers like Tolstoy and Proust "all these years." Our old literary habits "were just a side-effect of living in an environment of impoverished access."[28] Now that the Net has granted us abundant "access," Shirky concluded, we can at last lay those tired habits aside.

Such proclamations seem a little too staged to take seriously. They come off as the latest manifestation of the outré posturing that has always characterized the anti-intellectual wing of academia. But, then again, there may be a more charitable explanation. Federman, Shirky, and others like them may be early exemplars of the post-literary mind, intellectuals for whom the screen rather than the page has always been the primary conduit of information. As Alberto Manguel has written, "There is an unbridgeable chasm between the book that tradition has declared a classic and the book (the same book) that we have made ours through instinct, emotion and understanding: suffered through it, rejoiced in it, translated it into our experience and (notwithstanding the layers of readings with which a book comes into our hands) essentially become its first readers."[29] If you lack the time, the interest, or the facility to inhabit a literary work—to make it your own in the way Manguel describes—then of course you'd consider Tolstoy's masterpiece to be "too long, and not so interesting."

Although it may be tempting to ignore those who suggest the value of the literary mind has always been exaggerated, that would be a mistake. Their arguments are another important sign of the fundamental shift taking place in society's attitude toward intellectual achievement. Their words also make it a lot easier for people to justify that shift—to convince themselves that surfing the Web is a suitable, even superior, substitute for deep reading and other forms of calm and attentive thought. In arguing that books are archaic and dispensable, Federman and Shirky provide the intellectual cover that allows thoughtful people to slip comfortably into the permanent state of distractedness that defines the online life.

OUR DESIRE FOR fast-moving, kaleidoscopic diversions didn't originate with the invention of the World Wide Web. It has been present and growing for many decades, as the pace of our work and home lives has quickened and as broadcast media like radio and television have presented us with a welter of programs, messages, and advertisements. The Internet, though it marks a radical departure from traditional media in many ways, also represents a continuation of the intellectual and social trends that emerged from people's embrace of the electric media of the twentieth century and that have been shaping our lives and thoughts ever since. The distractions in our lives have been proliferating for a long time, but never has there been a medium that, like the Net, has been programmed to so widely scatter our attention and to do it so insistently.

David Levy, in *Scrolling Forward*, describes a meeting he attended at Xerox's famed Palo Alto Research Center in the mid-1970s, a time when the high-tech lab's engineers and programmers were devising many of the features we now take for granted in our personal computers. A group of prominent computer scientists had been invited to PARC to see a demonstration of a new operating system that made "multitasking" easy. Unlike traditional operating systems, which could display only one job at a time, the new system divided a screen into many "windows," each of which could run a different program or display a different document. To illustrate the flexibility of the system, the Xerox presenter clicked from a window in which he had been composing software code to another window that displayed a newly arrived e-mail message. He quickly read and replied to the message, then hopped back to the programming window and continued coding. Some in the audience applauded the new system. They saw that it would enable people to use their computers much more efficiently. Others recoiled from it. "Why in the world would you want to be interrupted—and distracted—by e-mail while programming?" one of the attending scientists angrily demanded.

The question seems quaint today. The windows interface has become the interface for all PCs and for most other computing devices as well. On the Net, there are windows within windows within windows, not to mention long ranks of tabs primed to trigger the opening of even more windows. Multitasking has become so routine that most of us would find it intolerable if we had to go back to computers that could run only one program or open only one file at a time. And yet, even though the question may have been rendered moot, it remains as vital today as it was thirty-five years ago. It points, as Levy says, to "a conflict between two different ways of working and two different understandings of how technology should be used to support that work." Whereas the Xerox researcher "was eager to juggle multiple threads of work simultaneously," the skeptical questioner viewed his own work "as an exercise in solitary, singleminded concentration."[30] In the choices we have made, consciously or not, about how we use our computers, we have rejected the intellectual tradition of solitary, singleminded concentration, the ethic that the book bestowed on us. We have cast our lot with the juggler.

Seven

THE JUGGLER'S BRAIN

I t's been a while since the first-person singular was heard in these pages. This seems like a good time for me, your word-processing scribe, to make a brief reappearance. I realize that I've dragged you through a lot of space and time over the last few chapters, and I appreciate your fortitude in sticking with me. The journey you've been on is the same one I took in trying to figure out what's been going on inside my head. The deeper I dug into the science of neuroplasticity and the progress of intellectual technology, the clearer it became that the Internet's import and influence can be judged only when viewed in the fuller context of intellectual history. As revolutionary as it may be, the Net is best understood as the latest in a long series of tools that have helped mold the human mind.

Now comes the crucial question: What can science tell us about the actual effects that Internet use is having on the way our minds work? No doubt, this question will be the subject of a great deal of research in the years ahead. Already, though, there is much we know or can surmise. The news is even more disturbing than I had suspected. Dozens of studies by psychologists, neurobiologists, educators, and Web designers point to the same conclusion: when we go online, we enter an environment that promotes cursory reading, hurried and distracted thinking, and superficial learning. It's possible to think deeply while surfing the Net, just as it's possible to think shallowly while reading a book, but that's not the type of thinking the technology encourages and rewards.

One thing is very clear: if, knowing what we know today about the brain's plasticity, you were to set out to invent a medium that would rewire our mental circuits as quickly and thoroughly as possible, you would probably end up designing something that looks and works a lot like the Internet. It's not just that we tend to use the Net regularly, even obsessively. It's that the Net delivers precisely the kind of sensory and cognitive stimuli—repetitive, intensive, interactive, addictive—that have been shown to result in strong and rapid alterations in brain circuits and functions. With the exception of alphabets and number systems, the Net may well be the single most powerful mind-altering technology that has ever come into general use. At the very least, it's the most powerful that has come along since the book.

During the course of a day, most of us with access to the Web spend at least a couple of hours online—sometimes much more—and during that time, we tend to repeat the same or similar actions over and over again, usually at a high rate of speed and often in response to cues delivered through a screen or a speaker. Some of the actions are physical ones. We tap the keys on our PC keyboard. We drag a mouse and click its left and right buttons and spin its scroll wheel. We draw the tips of our fingers across a trackpad. We use our thumbs to punch out text on the real or simulated keypads of our mobile phones. We rotate our iPhones and iPads to shift between "landscape" and "portrait" modes while manipulating the icons on their touch-sensitive screens.

As we go through these motions, the Net delivers a steady stream of inputs to our visual, somatosensory, and auditory cortices. There are the sensations that come through our hands and fingers as we click and scroll, type and touch. There are the many audio signals delivered through our ears, such as the chime that announces the arrival of a new e-mail or instant message and the various ringtones that our mobile phones use to alert us to different events. And, of course, there are the myriad visual cues that flash across our retinas as we navigate the online world: not just the ever-changing arrays of text and pictures and videos but also the hyperlinks distinguished by underlining or colored text, the cursors that change shape depending on their function, the new e-mail subject lines highlighted in bold type, the virtual buttons that call out to be clicked, the icons and other screen elements that beg to be dragged and dropped, the forms that require filling out, the pop-up ads and windows that need to be read or dismissed. The Net engages all of our senses—except, so far, those of smell and taste—and it engages them simultaneously.

The Net also provides a high-speed system for delivering responses and rewards—"positive reinforcements," in psychological terms—which encourage the repetition of both physical and mental actions. When we click a link, we get something new to look at and evaluate. When we Google a keyword, we receive, in the blink of an eye, a list of interesting information to appraise. When we send a text or an instant message or an e-mail, we often get a reply in a matter of seconds or minutes. When we use Facebook, we attract new friends or form closer bonds with old ones. When we send a tweet through Twitter, we gain new followers. When we write a blog post, we get comments from readers or links from other bloggers. The Net's interactivity gives us powerful new tools for finding information, expressing ourselves, and conversing with others. It also turns us into lab rats constantly pressing levers to get tiny pellets of social or intellectual nourishment.

The Net commands our attention with far greater insistency than our television or radio or morning newspaper ever did. Watch a kid texting his friends or a college student looking over the roll of new messages and requests on her Facebook page or a businessman scrolling through his e-mails on his phone—or consider yourself as you enter keywords into Google's search box and begin following a trail of links. What you see is a mind consumed with a medium. When we're online, we're often oblivious to everything else going on around us. The real world recedes as we process the flood of symbols and stimuli coming through our devices.

The interactivity of the Net amplifies this effect as well. Because we're often using our computers in a social context, to converse with friends or colleagues, to create "profiles" of ourselves, to broadcast our thoughts through blog posts or Facebook updates, our social standing is, in one way or another, always in play, always at risk. The resulting self-consciousness—even, at times, fear—magnifies the intensity of our involvement with the medium. That's true for everyone, but it's particularly true for the young, who tend to be compulsive in using their phones and computers for texting and messaging. Today's teenagers typically send or receive a message every few minutes throughout their waking hours. As the psychotherapist Michael Hausauer notes, teens and other young adults have a "terrific interest in knowing what's going on in the lives of their peers, coupled with a terrific anxiety about being out of the loop."[1] If they stop sending messages, they risk becoming invisible.

Our use of the Internet involves many paradoxes, but the one that promises to have the greatest long-term influence over how we think is this one: the Net seizes our attention only to scatter it. We focus intensively on the medium itself, on the flickering screen, but we're distracted by the medium's rapid-fire delivery of competing messages and stimuli. Whenever and wherever we log on, the Net presents us with an incredibly seductive blur. Human beings "want more information, more impressions, and more complexity," writes Torkel Klingberg, the Swedish neuroscientist. We tend to "seek out situations that demand concurrent performance or situations in which [we] are overwhelmed with information."[2] If the slow progression of words across printed pages dampened our craving to be inundated by mental stimulation, the Net indulges it. It returns us to our native state of bottom-up distractedness, while presenting us with far more distractions than our ancestors ever had to contend with.

Not all distractions are bad. As most of us know from experience, if we concentrate too intensively on a tough problem, we can get stuck in a mental rut. Our thinking narrows, and we struggle vainly to come up with new ideas. But if we let the problem sit unattended for a time—if we "sleep on it"—we often return to it with a fresh perspective and a burst of creativity. Research by Ap Dijksterhuis, a Dutch psychologist who heads the Unconscious Lab at Radboud University in Nijmegen, indicates that such breaks in our attention give our unconscious mind time to grapple with a problem, bringing to bear information and cognitive processes unavailable to conscious deliberation. We usually make better decisions, his experiments reveal, if we shift our attention away from a difficult mental challenge for a time. But Dijksterhuis's work also shows that our unconscious thought processes don't engage with a problem until we've clearly and consciously defined the problem.[3] If we don't have a particular intellectual goal in mind, Dijksterhuis writes, "unconscious thought does not occur."[4]

The constant distractedness that the Net encourages—the state of being, to borrow another phrase from Eliot's *Four Quartets*, "distracted from distraction by distraction"—is very different from the kind of temporary, purposeful diversion of our mind that refreshes our thinking when we're weighing a decision. The Net's cacophony of stimuli short-circuits both conscious and unconscious thought, preventing our minds from thinking either deeply or creatively. Our brains turn into simple signal-processing units, quickly shepherding information into consciousness and then back out again.

In a 2005 interview, Michael Merzenich ruminated on the Internet's power to cause not just modest alterations but fundamental changes in our mental makeup. Noting that "our brain is modified on a substantial scale, physically and functionally, each time we learn a new skill or develop a new ability," he described the Net as the latest in a series of "modern cultural specializations" that "contemporary humans can spend millions of 'practice' events at [and that] the average human a thousand years ago had absolutely no exposure to." He concluded that "our brains are massively remodeled by this exposure."[5] He returned to this theme in a post on his blog in 2008, resorting to capital letters to emphasize his points. "When culture drives changes in the ways that we engage our brains, it creates DIFFERENT brains," he wrote, noting that our minds "strengthen specific heavily-exercised processes." While acknowledging that it's now hard to imagine living without the Internet and online tools like the Google search engine, he stressed that "THEIR HEAVY USE HAS NEUROLOGICAL CONSEQUENCES."[6]

What we're *not* doing when we're online also has neurological consequences. Just as neurons that fire together wire together, neurons that don't fire together don't wire together. As the time we spend scanning Web pages crowds out the time we spend reading books, as the time we spend exchanging bite-sized text messages crowds out the time we spend composing sentences and paragraphs, as the time we spend hopping across links crowds out the time we devote to quiet reflection and contemplation, the circuits that support those old intellectual functions and pursuits weaken and begin to break apart. The brain recycles the disused neurons and synapses for other, more pressing work. We gain new skills and perspectives but lose old ones.

GARY SMALL, A professor of psychiatry at UCLA and the director of its Memory and Aging Center, has been studying the physiological and neurological effects of the use of digital media, and what he's discovered backs up Merzenich's belief that the Net causes extensive brain changes. "The current explosion of digital technology not only is changing the way we live and communicate but is rapidly and profoundly altering our brains," he says. The daily use of computers, smartphones, search engines, and other such tools "stimulates brain cell alteration and neurotransmitter release, gradually strengthening new neural pathways in our brains while weakening old ones."[7]

In 2008, Small and two of his colleagues carried out the first experiment that actually showed people's brains changing in response to Internet use.[8] The researchers recruited twenty-four volunteers—a dozen experienced Web surfers and a dozen novices—and scanned their brains as they performed searches on Google. (Since a computer won't fit inside a magnetic resonance imager, the subjects were equipped with goggles onto which were projected images of Web pages, along with a small handheld touchpad to navigate the pages.) The scans revealed that the brain activity of the experienced Googlers was much broader than that of the novices. In particular, "the computer-savvy subjects used a specific network in the left front part of the brain, known as the dorsolateral prefrontal cortex, [while] the Internet-naïve subjects showed minimal, if any, activity in this area." As a control for the test, the researchers also had the subjects read straight text in a simulation of book reading; in this case, scans revealed no significant difference in brain activity between the two groups. Clearly, the experienced Net users' distinctive neural pathways had developed through their Internet use.

The most remarkable part of the experiment came when the tests were repeated six days later. In the interim, the researchers had the novices spend an hour a day online, searching the Net. The new scans revealed that the area in their prefrontal cortex that had been largely dormant now showed extensive activity—just like the activity in the brains of the veteran surfers. "After just five days of practice, the exact same neural circuitry in the front part of the brain became active in the Internet-naïve subjects," reports Small. "Five hours on the Internet, and the naïve subjects had already rewired their brains." He goes on to ask, "If our brains are so sensitive to just an hour a day of computer exposure, what happens when we spend more time [online]?"[9]

One other finding of the study sheds light on the differences between reading Web pages and reading books. The researchers found that when people search the Net they exhibit a very different pattern of brain activity than they do when they read book-like text. Book readers have a lot of activity in regions associated with language, memory, and visual processing, but they don't display much activity in the prefrontal regions associated with decision making and problem solving. Experienced Net users, by contrast, display extensive activity across all those brain regions when they scan and search Web pages. The good news here is that Web surfing, because it engages so many brain functions, may help keep older people's minds sharp. Searching and browsing seem to "exercise" the brain in a way similar to solving crossword puzzles, says Small.

But the extensive activity in the brains of surfers also points to why deep reading and other acts of sustained concentration become so difficult online. The need to evaluate links and make related navigational choices, while also processing a multiplicity of fleeting sensory stimuli, requires constant mental coordination and decision making, distracting the brain from the work of interpreting text or other information. Whenever we, as readers, come upon a link, we have to pause, for at least a split second, to allow our prefrontal cortex to evaluate whether or not we should click on it. The redirection of our mental resources, from reading words to making judgments, may be imperceptible to us—our brains are quick—but it's been shown to impede comprehension and retention, particularly when it's repeated frequently. As the executive functions of the prefrontal cortex kick in, our brains become not only exercised but overtaxed. In a very real way, the Web returns us to the time of *scriptura continua*, when reading was a cognitively strenuous act. In reading online, Maryanne Wolf says, we sacrifice the facility that makes deep reading possible. We revert to being "mere decoders of information."[10] Our ability to make the rich mental connections that form when we read deeply and without distraction remains largely disengaged.

Steven Johnson, in his 2005 book *Everything Bad Is Good for You*, contrasted the widespread, teeming neural activity seen in the brains of computer users with the much more muted activity evident in the brains of book readers. The comparison led him to suggest that computer use provides more intense mental stimulation than does book reading. The neural evidence could even, he wrote, lead a person to conclude that "reading books chronically understimulates the senses."[11] But while Johnson's diagnosis is correct, his interpretation of the differing patterns of brain activity is misleading. It is the very fact that book reading "understimulates the senses" that makes the activity so intellectually rewarding. By allowing us to filter out distractions, to quiet the problem-solving functions of the frontal lobes, deep reading becomes a form of deep thinking. The mind of the experienced book reader is a calm mind, not a buzzing one. When it comes to the firing of our neurons, it's a mistake to assume that more is better.

John Sweller, an Australian educational psychologist, has spent three decades studying how our minds process information and, in particular, how we learn. His work illuminates how the Net and other media influence the style and the depth of our thinking. Our brains, he explains, incorporate two very different kinds of memory: short-term and long-term. We hold our immediate impressions, sensations, and thoughts as short-term memories, which tend to last only a matter of seconds. All the things we've learned about the world, whether consciously or unconsciously, are stored as long-term memories, which can remain in our brains for a few days, a few years, or even a lifetime. One particular type of short-term memory, called working memory, plays an instrumental role in the transfer of information into long-term memory and hence in the creation of our personal store of knowledge. Working memory forms, in a very real sense, the contents of our consciousness at any given moment. "We are conscious of what is in working memory and not conscious of anything else," says Sweller.[12]

If working memory is the mind's scratch pad, then long-term memory is its filing system. The contents of our long-term memory lie mainly outside of our consciousness. In order for us to think about something we've previously learned or experienced, our brain has to transfer the memory from long-term memory back into working memory. "We are only aware that something was stored in long-term memory when it is brought down into working memory," explains Sweller.[13] It was once assumed that long-term memory served merely as a big warehouse of facts, impressions, and events, that it "played little part in complex cognitive processes such as thinking and problem-solving."[14] But brain scientists have come to realize that long-term memory is actually the seat of understanding. It stores not just facts but complex concepts, or "schemas." By organizing scattered bits of information into patterns of knowledge, schemas give depth and richness to our thinking. "Our intellectual prowess is derived largely from the schemas we have acquired over long periods of time," says Sweller. "We are able to understand concepts in our areas of expertise because we have schemas associated with those concepts."[15]

The depth of our intelligence hinges on our ability to transfer information from working memory to long-term memory and weave it into conceptual schemas. But the passage from working memory to long-term memory also forms the major bottleneck in our brain. Unlike long-term memory, which has a vast capacity, working memory is able to hold only a very small amount of information. In a renowned 1956 paper, "The Magical Number Seven, Plus or Minus Two," Princeton psychologist George Miller observed that working memory could typically hold just seven pieces, or "elements," of information. Even that is now considered an overstatement. According to Sweller, current evidence suggests that "we can process no more than about two to four elements at any given time with the actual number probably being at the lower [rather] than the higher end of this scale." Those elements that we are able to hold in working memory will, moreover, quickly vanish "unless we are able to refresh them by rehearsal."[16]

Imagine filling a bathtub with a thimble; that's the challenge involved in transferring information from working memory into long-term memory. By regulating the velocity and intensity of information flow, media exert a strong influence on this process. When we read a book, the information faucet provides a steady drip, which we can control by the pace of our reading. Through our single-minded concentration on the text, we can transfer all or most of the information, thimbleful by thimbleful, into long-term memory and forge the rich associations essential to the creation of schemas. With the Net, we face many information faucets, all going full blast. Our little thimble overflows as we rush from one faucet to the next. We're able to transfer only a small portion of the information to long-term memory, and what we do transfer is a jumble of drops from different faucets, not a continuous, coherent stream from one source.

The information flowing into our working memory at any given moment is called our "cognitive load." When the load exceeds our mind's ability to store and process the information—when the water overflows the thimble—we're unable to retain the information or to draw connections with the information already stored in our long-term memory. We can't translate the new information into schemas. Our ability to learn suffers, and our understanding remains shallow. Because our ability to maintain our attention also depends on our working memory—"we have to remember what it is we are to concentrate on," as Torkel Klingberg says—a high cognitive load amplifies the distractedness we experience. When our brain is overtaxed, we find "distractions more distracting."[17] (Some studies link attention deficit disorder, or ADD, to the overloading of working memory.) Experiments indicate that as we reach the limits of our working memory, it becomes harder to distinguish relevant information from irrelevant information, signal from noise. We become mindless consumers of data.

Difficulties in developing an understanding of a subject or a concept appear to be "heavily determined by working memory load," writes Sweller, and the more complex the material we're trying to learn, the greater the penalty exacted by an overloaded mind.[18] There are many possible sources of cognitive overload, but two of the most important, according to Sweller, are "extraneous problem-solving" and "divided attention." Those also happen to be two of the central features of the Net as an informational medium. Using the Net may, as Gary Small suggests, exercise the brain the way solving crossword puzzles does. But such intensive exercise, when it becomes our primary mode of thought, can impede deep learning and thinking. Try reading a book while doing a crossword puzzle; that's the intellectual environment of the Internet.

BACK IN THE 1980s, when schools began investing heavily in computers, there was much enthusiasm about the apparent advantages of digital documents over paper ones. Many educators were convinced that introducing hyperlinks into text displayed on computer screens would be a boon to learning. Hypertext would, they argued, strengthen students' critical thinking by enabling them to switch easily between different viewpoints. Freed from the lockstep reading demanded by printed pages, readers would make all sorts of new intellectual connections among diverse texts. The academic enthusiasm for hypertext was further kindled by the belief, in line with the fashionable postmodern theories of the day, that hypertext would overthrow the patriarchal authority of the author and shift power to the reader. It would be a technology of liberation. Hypertext, wrote the literary theorists George Landow and Paul Delany, can "provide a revelation" by freeing readers from the "stubborn materiality" of printed text. By "moving away from the constrictions of page-bound technology," it "provides a better model for the mind's ability to reorder the elements of experience by changing the links of association or determination between them."[19]

By the end of the decade, the enthusiasm had begun to subside. Research was painting a fuller, and very different, picture of the cognitive effects of hypertext. Evaluating links and navigating a path through them, it turned out, involves mentally demanding problem-solving tasks that are extraneous to the act of reading itself. Deciphering hypertext substantially increases readers' cognitive load and hence weakens their ability to comprehend and retain what they're reading. A 1989 study showed that readers of hypertext often ended up clicking distractedly "through pages instead of reading them carefully." A 1990 experiment revealed that hypertext readers often "could not remember what they had and had not read." In another study that same year, researchers had two groups of people answer a series of questions by searching through a set of documents. One group searched through electronic hypertext documents, while the other searched through traditional paper documents. The group that used the paper documents outperformed the hypertext group in completing the assignment. In reviewing the results of these and other experiments, the editors of a 1996 book on hypertext and cognition wrote that, since hypertext "imposes a higher cognitive load on the reader," it's no surprise "that empirical comparisons between paper presentation (a familiar situation) and hypertext (a new, cognitively demanding situation) do not always favor hypertext." But they predicted that, as readers gained greater "hypertext literacy," the cognition problems would likely diminish.[20]

That hasn't happened. Even though the World Wide Web has made hypertext commonplace, indeed ubiquitous, research continues to show that people who read linear text comprehend more, remember more, and learn more than those who read text peppered with links. In a 2001 study, two Canadian scholars asked seventy people to read "The Demon Lover," a short story by the modernist writer Elizabeth Bowen. One group read the story in a traditional linear-text format; a second group read a version with links, as you'd find on a Web page. The hypertext readers took longer to read the story, yet in subsequent interviews they also reported more confusion and uncertainty about what they had read. Three-quarters of them said that they had difficulty following the text, while only one in ten of the linear-text readers reported such problems. One hypertext reader complained, "The story was very jumpy. I don't know if that was caused by the hypertext, but I made choices and all of a sudden it wasn't flowing properly, it just kind of jumped to a new idea I didn't really follow."

A second test by the same researchers, using a shorter and more simply written story, Sean O'Faolain's "The Trout," produced the same results. Hypertext readers again reported greater confusion following the text, and their comments about the story's plot and imagery were less detailed and less precise than those of the linear-text readers. With hypertext, the researchers concluded, "the absorbed and personal mode of reading seems to be discouraged." The readers' attention "was directed toward the machinery of the hypertext and its functions rather than to the experience offered by the story."[21] The medium used to present the words obscured the meaning of the words.

In another experiment, researchers had people sit at computers and review two online articles describing opposing theories of learning. One article laid out an argument that "knowledge is objective"; the other made the case that "knowledge is relative." Each article was set up in the same way, with similar headings, and each had links to the other article, allowing a reader to jump quickly between the two to compare the theories. The researchers hypothesized that people who used the links would gain a richer understanding of the two theories and their differences than would people who read the pages sequentially, completing one before going on to the other. They were wrong. The test subjects who read the pages linearly actually scored considerably higher on a subsequent comprehension test than those who clicked back and forth between the pages. The links got in the way of learning, the researchers concluded.[22]

Another researcher, Erping Zhu, conducted a different kind of experiment that was also aimed at discerning the influence of hypertext on comprehension. She had groups of people read the same piece of online writing, but she varied the number of links included in the passage. She then tested the readers' comprehension by asking them to write a summary of what they had read and complete a multiple-choice test. She found that comprehension declined as the number of links increased. Readers were forced to devote more and more of their attention and brain power to evaluating the links and deciding whether to click on them. That left less attention and fewer cognitive resources to devote to understanding what they were reading. The experiment suggested a strong correlation "between the number of links and disorientation or cognitive over-load," wrote Zhu. "Reading and comprehension require establishing relationships between concepts, drawing inferences, activating prior knowledge, and synthesizing main ideas. Disorientation or cognitive overload may thus interfere with cognitive activities of reading and comprehension."[23]

In 2005, Diana DeStefano and Jo-Anne LeFevre, psychologists with the Centre for Applied Cognitive Research at Canada's Carleton University, undertook a comprehensive review of thirty-eight past experiments involving the reading of hypertext. Although not all the studies showed that hypertext diminished comprehension, they found "very little support" for the once-popular theory "that hyper-text will lead to an enriched experience of the text." To the con-trary, the preponderance of evidence indicated that "the increased demands of decision-making and visual processing in hypertext impaired reading performance," particularly when compared to "traditional linear presentation." They concluded that "many fea-tures of hypertext resulted in increased cognitive load and thus may have required working memory capacity that exceeded readers' capabilities."[24]

THE WEB COMBINES the technology of hypertext with the tech-nology of multimedia to deliver what's called "hypermedia." It's not just words that are served up and electronically linked, but also images, sounds, and moving pictures. Just as the pioneers of hypertext once believed that links would provide a richer learn-ing experience for readers, many educators also assumed that mul-timedia, or "rich media," as it's sometimes called, would deepen comprehension and strengthen learning. The more inputs, the bet-ter. But this assumption, long accepted without much evidence, has also been contradicted by research. The division of attention demanded by multimedia further strains our cognitive abilities, diminishing our learning and weakening our understanding. When

it comes to supplying the mind with the stuff of thought, more
can be less.

In a study published in the journal *Media Psychology* in 2007,
researchers recruited more than a hundred volunteers to watch
a presentation about the country of Mali played through a Web
browser on a computer. Some of the subjects watched a version of
the presentation that included only a series of text pages. Another
group watched a version that included, along with the pages of text, a
window in which an audiovisual presentation of related material was
streamed. The test subjects were able to stop and start the stream as
they wished.

After viewing the presentation, the subjects took a ten-question
quiz on the material. The text-only viewers answered an average
of 7.04 of the questions correctly, while the multimedia viewers
answered just 5.98 correctly—a significant difference, according to
the researchers. The subjects were also asked a series of questions
about their perceptions of the presentation. The text-only readers
found it to be more interesting, more educational, more understandable, and more enjoyable than did the multimedia viewers, and the
multimedia viewers were much more likely to agree with the statement "I did not learn anything from this presentation" than were the
text-only readers. The multimedia technologies so common to the
Web, the researchers concluded, "would seem to limit, rather than
enhance, information acquisition."[25]

In another experiment, a pair of Cornell researchers divided a
class of students into two groups. One group was allowed to surf the
Web while listening to a lecture. A log of their activity showed that
they looked at sites related to the lecture's content but also visited
unrelated sites, checked their e-mail, went shopping, watched videos, and did all the other things that people do online. The second
group heard the identical lecture but had to keep their laptops shut.
Immediately afterward, both groups took a test measuring how well
they could recall the information from the lecture. The surfers, the
researchers report, "performed significantly poorer on immediate

measures of memory for the to-be-learned content." It didn't matter, moreover, whether they surfed information related to the lecture or completely unrelated content—they all performed poorly. When the researchers repeated the experiment with another class, the results were the same.[26]

Kansas State University scholars conducted a similarly realistic study. They had a group of college students watch a typical CNN broadcast in which an anchor reported four news stories while various info-graphics flashed on the screen and a textual news crawl ran along the bottom. They had a second group watch the same programming but with the graphics and the news crawl stripped out. Subsequent tests found that the students who had watched the multimedia version remembered significantly fewer facts from the stories than those who had watched the simpler version. "It appears," wrote the researchers, "that this multimessage format exceeded viewers' attentional capacity."[27]

Supplying information in more than one form doesn't always take a toll on understanding. As we all know from reading illustrated textbooks and manuals, pictures can help clarify and reinforce written explanations. Education researchers have also found that carefully designed presentations that combine audio and visual explanations or instructions can enhance students' learning. The reason, current theories suggest, is that our brains use different channels for processing what we see and what we hear. As Sweller explains, "Auditory and visual working memory are separate, at least to some extent, and because they are separate, effective working memory may be increased by using both processors rather than one." As a result, in some cases "the negative effects of split attention might be ameliorated by using both auditory and visual modalities"—sounds and pictures, in other words.[28] The Internet, however, wasn't built by educators to optimize learning. It presents information not in a carefully balanced way but as a concentration-fragmenting mishmash.

The Net is, by design, an interruption system, a machine geared for dividing attention. That's not only a result of its ability to display

many different kinds of media simultaneously. It's also a result of
the ease with which it can be programmed to send and receive messages. Most e-mail applications, to take an obvious example, are set
up to check automatically for new messages every five or ten minutes, and people routinely click the "check for new mail" button
even more frequently than that. Studies of office workers who use
computers reveal that they constantly stop what they're doing to read
and respond to incoming e-mails. It's not unusual for them to glance
at their in-box thirty or forty times an hour (though when asked how
frequently they look, they'll often give a much lower figure).[29] Since
each glance represents a small interruption of thought, a momentary
redeployment of mental resources, the cognitive cost can be high.
Psychological research long ago proved what most of us know from
experience: frequent interruptions scatter our thoughts, weaken our
memory, and make us tense and anxious. The more complex the
train of thought we're involved in, the greater the impairment the
distractions cause.[30]

Beyond the influx of personal messages—not only e-mail but also
instant messages and text messages—the Web increasingly supplies
us with all manner of other automated notifications. Feed readers
and news aggregators let us know whenever a new story appears at
a favorite publication or blog. Social networks alert us to what our
friends are doing, often moment by moment. Twitter and other
microblogging services tell us whenever one of the people we "follow" broadcasts a new message. We can also set up alerts to monitor
shifts in the value of our investments, news reports about particular people or events, updates to the software we use, new videos
uploaded to YouTube, and so forth. Depending on how many information streams we subscribe to and the frequency with which they
send out updates, we may field a dozen alerts an hour, and for the
most connected among us, the number can be much higher. Each of
them is a distraction, another intrusion on our thoughts, another bit
of information that takes up precious space in our working memory.

Navigating the Web requires a particularly intensive form of mental multitasking. In addition to flooding our working memory with information, the juggling imposes what brain scientists call "switching costs" on our cognition. Every time we shift our attention, our brain has to reorient itself, further taxing our mental resources. As Maggie Jackson explains in *Distracted*, her book on multitasking, "the brain takes time to change goals, remember the rules needed for the new task, and block out cognitive interference from the previous, still-vivid activity."[31] Many studies have shown that switching between just two tasks can add substantially to our cognitive load, impeding our thinking and increasing the likelihood that we'll overlook or misinterpret important information. In one simple experiment, a group of adults was shown a series of colored shapes and asked to make predictions based on what they saw. They had to perform the task while wearing headphones that played a series of beeps. In one trial, they were told to ignore the beeps and just concentrate on the shapes. In a second trial, using a different set of visual cues, they were told to keep track of the number of beeps. After each go-through, they completed a test that required them to interpret what they had just done. In both trials, the subjects made predictions with equal success. But after the multitasking trial, they had a much harder time drawing conclusions about their experience. Switching between the two tasks short-circuited their understanding; they got the job done, but they lost its meaning. "Our results suggest that learning facts and concepts will be worse if you learn them while you're distracted," said the lead researcher, UCLA psychologist Russell Poldrack.[32] On the Net, where we routinely juggle not just two but several mental tasks, the switching costs are all the higher.

It's important to emphasize that the Net's ability to monitor events and automatically send out messages and notifications is one of its great strengths as a communication technology. We rely on that capability to personalize the workings of the system, to program the vast database to respond to our particular needs, interests, and desires. We *want* to be interrupted, because each interruption brings us a valuable piece of information. To turn off these alerts is to risk feeling out of touch, or even socially isolated. The near-continuous stream of new information pumped out by the Web also plays to our natural tendency to "vastly overvalue what happens to us *right now*," as Union College psychologist Christopher Chabris explains. We crave the new even when we know that "the new is more often trivial than essential."[33]

And so we ask the Internet to keep interrupting us, in ever more and different ways. We willingly accept the loss of concentration and focus, the division of our attention and the fragmentation of our thoughts, in return for the wealth of compelling or at least divert-ing information we receive. Tuning out is not an option many of us would consider.

IN 1879, A French ophthalmologist named Louis Émile Javal discov-ered that when people read, their eyes don't sweep across the words in a perfectly fluid way. Their visual focus advances in little jumps, called saccades, pausing briefly at different points along each line. One of Javal's colleagues at the University of Paris soon made another dis-covery: that the pattern of pauses, or "eye fixations," can vary greatly depending on what's being read and who's doing the reading. In the wake of these discoveries, brain researchers began to use eye-tracking experiments to learn more about how we read and how our minds work. Such studies have also proven valuable in providing further insights into the Net's effects on attention and cognition.

In 2006, Jakob Nielsen, a longtime consultant on the design of Web pages who has been studying online reading since the 1990s, conducted an eye-tracking study of Web users. He had 232 people wear a small camera that tracked their eye movements as they read pages of text and browsed other content. Nielsen found that hardly any of the participants read online text in a methodical, line-by-line way, as they'd typically read a page of text in a book. The vast major-ity skimmed the text quickly, their eyes skipping down the page in a pattern that resembled, roughly, the letter *F*. They'd start by glanc-

ing all the way across the first two or three lines of text. Then their
eyes would drop down a bit, and they'd scan about halfway across a
few more lines. Finally, they'd let their eyes cursorily drift a little
farther down the left-hand side of the page. This pattern of online
reading was confirmed by a subsequent eye-tracking study carried
out at the Software Usability Research Laboratory at Wichita State
University.[34]

"F," wrote Nielsen, in summing up the findings for his clients,
is "for *fast*. That's how users read your precious content. In a few
seconds, their eyes move at amazing speeds across your website's
words in a pattern that's very different from what you learned in
school."[35] As a complement to his eye-tracking study, Nielsen analyzed an extensive database on the behavior of Web users that had
been compiled by a team of German researchers. They had monitored the computers of twenty-five people for an average of about
a hundred days each, tracking the time the subjects spent looking
at some fifty thousand Web pages. Parsing the data, Nielsen found
that as the number of words on a page increases, the time a visitor spends looking at the page goes up, but only slightly. For every
hundred additional words, the average viewer will spend just 4.4
more seconds perusing the page. Since even the most accomplished
reader can read only about eighteen words in 4.4 seconds, Nielsen
told his clients, "when you add verbiage to a page, you can assume
that customers will read 18% of it." And that, he cautioned, is almost
certainly an overstatement. It's unlikely that the people in the study
were spending all their time reading; they were also probably glancing at pictures, videos, advertisements, and other types of content.[36]

Nielsen's analysis backed up the conclusions of the German
researchers themselves. They had reported that most Web pages are
viewed for ten seconds or less. Fewer than one in ten page views
extend beyond two minutes, and a significant portion of those
seem to involve "unattended browser windows . . . left open in the
background of the desktop." The researchers observed that "even
new pages with plentiful information and many links are regularly

viewed only for a brief period." The results, they said, "confirm that browsing is a rapidly interactive activity."[37] The results also reinforce something that Nielsen wrote in 1997 after his first study of online reading. "How do users read on the web?" he asked then. His succinct answer: "They don't."[38]

Web sites routinely collect detailed data on visitor behavior, and those statistics underscore just how quickly we leap between pages when we're online. Over a period of two months in 2008, an Israeli company named ClickTale, which supplies software for analyzing how people use corporate Web pages, collected data on the behavior of a million visitors to sites maintained by its clients around the world. It found that in most countries people spend, on average, between nineteen and twenty-seven seconds looking at a page before moving on to the next one, including the time required for the page to load into their browser's window. German and Canadian surfers spend about twenty seconds on each page, U.S. and U.K. surfers spend about twenty-one seconds, Indians and Australians spend about twenty-four seconds, and the French spend about twenty-five seconds.[39] On the Web, there is no such thing as leisurely browsing. We want to gather as much information as quickly as our eyes and fingers can move.

That's true even when it comes to academic research. As part of a five-year study that ended in early 2008, a group from University College London examined computer logs documenting the behavior of visitors to two popular research sites, one operated by the British Library and one by a U.K. educational consortium. Both sites provided users with access to journal articles, e-books, and other sources of written information. The scholars found that people using the sites exhibited a distinctive "form of skimming activity" in which they'd hop quickly from one source to another, rarely returning to any source they had already visited. They'd typically read, at most, one or two pages of an article or book before "bouncing out" to another site. "It is clear that users are not reading online in the traditional sense," the authors of the study reported; "indeed there are signs that new forms of 'reading' are emerging as users 'power browse' horizontally through titles, contents pages and abstracts going for quick wins. It almost seems that they go online to avoid reading in the traditional sense."[40]

The shift in our approach to reading and research seems to be an inevitable consequence of our reliance on the technology of the Net, argues Merzenich, and it bespeaks a deeper change in our thinking. "There is absolutely no question that modern search engines and cross-referenced websites have powerfully enabled research and communication efficiencies," he says. "There is also absolutely no question that our brains are engaged less directly and more shallowly in the synthesis of information when we use research strategies that are all about 'efficiency,' 'secondary (and out-of-context) referencing,' and 'once over, lightly.'"[41]

The switch from reading to power-browsing is happening very quickly. Already, reports Ziming Liu, a library science professor at San José State University, "the advent of digital media and the growing collection of digital documents have had a profound impact on reading." In 2003, Liu surveyed 113 well-educated people— engineers, scientists, accountants, teachers, business managers, and graduate students, mainly between thirty and forty-five years old—to gauge how their reading habits had changed over the preceding ten years. Nearly eighty-five percent of the people reported that they were spending more time reading electronic documents. When asked to characterize how their reading practices have changed, eighty-one percent said that they were spending more time "browsing and scanning," and eighty-two percent reported that they were doing more "non-linear reading." Only twenty-seven percent said that the time they devoted to "in-depth reading" was on the rise, while forty-five percent said it was declining. Just sixteen percent said they were giving more "sustained attention" to reading; fifty percent said they were giving it less "sustained attention."

The findings, said Liu, indicate that "the digital environment tends to encourage people to explore many topics extensively, but at a more superficial level," and that "hyperlinks distract people from reading and thinking deeply." One of the participants in the study told Liu, "I find that my patience with reading long documents is decreasing. I want to skip ahead to the end of long articles." Another said, "I skim much more [when reading] html pages than I do with printed materials." It's quite clear, Liu concluded, that with the flood of digital text pouring through our computers and phones, "people are spending more time on reading" than they used to. But it's equally clear that it's a very different kind of reading. A "screen-based reading behavior is emerging," he wrote, which is characterized by "browsing and scanning, keyword spotting, one-time reading, [and] non-linear reading." The time "spent on in-depth reading and concentrated reading" is, on the other hand, falling steadily.[42]

There's nothing wrong with browsing and scanning, or even power-browsing and power-scanning. We've always skimmed newspapers more than we've read them, and we routinely run our eyes over books and magazines in order to get the gist of a piece of writing and decide whether it warrants more thorough reading. The ability to skim text is every bit as important as the ability to read deeply. What is different, and troubling, is that skimming is becoming our dominant mode of reading. Once a means to an end, a way to identify information for deeper study, scanning is becoming an end in itself—our preferred way of gathering and making sense of information of all sorts. We've reached the point where a Rhodes Scholar like Florida State's Joe O'Shea—a philosophy major, no less—is comfortable admitting not only that he doesn't read books but that he doesn't see any particular need to read them. Why bother, when you can Google the bits and pieces you need in a fraction of a second? What we're experiencing is, in a metaphorical sense, a reversal of the early trajectory of civilization: we are evolving from being cultivators of personal knowledge to being hunters and gatherers in the electronic data forest.

THERE ARE COMPENSATIONS. Research shows that certain cognitive skills are strengthened, sometimes substantially, by our use of computers and the Net. These tend to involve lower-level, or more primitive, mental functions such as hand-eye coordination, reflex response, and the processing of visual cues. One much-cited study of video gaming, published in *Nature* in 2003, revealed that after just ten days of playing action games on computers, a group of young people had significantly increased the speed with which they could shift their visual focus among different images and tasks. Veteran game players were also found to be able to identify more items in their visual field than novices could. The authors of the study concluded that "although video-game playing may seem to be rather mindless, it is capable of radically altering visual attentional processing."[43]

While experimental evidence is sparse, it seems only logical that Web searching and browsing would also strengthen brain functions related to certain kinds of fast-paced problem solving, particularly those involving the recognition of patterns in a welter of data. Through the repetitive evaluation of links, headlines, text snippets, and images, we should become more adept at quickly distinguishing among competing informational cues, analyzing their salient characteristics, and judging whether they'll have practical benefit for whatever task we're engaged in or goal we're pursuing. One British study of the way women search for medical information online indicated that the speed with which they were able to assess the probable value of a Web page increased as they gained familiarity with the Net.[44] It took an experienced browser only a few seconds to make an accurate judgment about whether a page was likely to have trustworthy information.

Other studies suggest that the kind of mental calisthenics we engage in online may lead to a small expansion in the capacity of our working memory.[45] That, too, would help us to become more adept at juggling data. Such research "indicates that our brains learn to swiftly focus attention, analyze information, and almost instantaneously decide on a go or no-go decision," says Gary Small. He believes that as we spend more time navigating the vast quantity of information available online, "many of us are developing neural circuitry that is customized for rapid and incisive spurts of directed attention."[46] As we practice browsing, surfing, scanning, and multitasking, our plastic brains may well become more facile at those tasks.

The importance of such skills shouldn't be taken lightly. As our work and social lives come to center on the use of electronic media, the faster we're able to navigate those media and the more adroitly we're able to shift our attention among online tasks, the more valuable we're likely to become as employees and even as friends and colleagues. As the writer Sam Anderson put it in "In Defense of Distraction," a 2009 article in *New York* magazine, "Our jobs depend on connectivity" and "our pleasure-cycles—no trivial matter—are increasingly tied to it." The practical benefits of Web use are many, which is one of the main reasons we spend so much time online. "It's too late," argues Anderson, "to just retreat to a quieter time."[47]

He's right, but it would be a serious mistake to look narrowly at the Net's benefits and conclude that the technology is making us more intelligent. Jordan Grafman, head of the cognitive neuroscience unit at the National Institute of Neurological Disorders and Stroke, explains that the constant shifting of our attention when we're online may make our brains more nimble when it comes to multitasking, but improving our ability to multitask actually hampers our ability to think deeply and creatively. "Does optimizing for multitasking result in better functioning—that is, creativity, inventiveness, productiveness? The answer is, in more cases than not, no," says Grafman. "The more you multitask, the less deliberative you become; the less able to think and reason out a problem." You become, he argues, more likely to rely on conventional ideas and solutions rather than challenging them with original lines of thought.[48] David Meyer, a University of Michigan neuroscientist and one of the leading experts on multitasking, makes a similar point. As we gain more experience in rapidly shifting our attention, we may "overcome some of the inefficiencies" inherent in multitasking, he says, "but except in rare circumstances, you can train until you're blue in the face and you'd never be as good as if you just focused on one thing at a time."[49] What we're doing when we multitask "is learning to be skillful at a superficial level."[50] The Roman philosopher Seneca may have put it best two thousand years ago: "To be everywhere is to be nowhere."[51]

In an article published in *Science* in early 2009, Patricia Greenfield, a prominent developmental psychologist who teaches at UCLA, reviewed more than fifty studies of the effects of different types of media on people's intelligence and learning ability. She concluded that "every medium develops some cognitive skills at the expense of others." Our growing use of the Net and other screen-based technologies has led to the "widespread and sophisticated development of visual-spatial skills." We can, for example, rotate objects in our minds better than we used to be able to. But our "new strengths in visual-spatial intelligence" go hand in hand with a weakening of our capacities for the kind of "deep processing" that underpins "mindful knowledge acquisition, inductive analysis, critical thinking, imagination, and reflection."[52] The Net is making us smarter, in other words, only if we define intelligence by the Net's own standards. If we take a broader and more traditional view of intelligence—if we think about the depth of our thought rather than just its speed—we have to come to a different and considerably darker conclusion.

Given our brain's plasticity, we know that our online habits continue to reverberate in the workings of our synapses when we're not online. We can assume that the neural circuits devoted to scanning, skimming, and multitasking are expanding and strengthening, while those used for reading and thinking deeply, with sustained concentration, are weakening or eroding. In 2009, researchers from Stanford University found signs that this shift may already be well under way. They gave a battery of cognitive tests to a group of heavy media multitaskers as well as a group of relatively light multitaskers. They found that the heavy multitaskers were much more easily distracted by "irrelevant environmental stimuli," had significantly less control over the contents of their working memory, and were in general much less able to maintain their concentration on a particular task. Whereas the infrequent multitaskers exhibited relatively strong "top-down attentional control," the habitual multitaskers showed "a greater tendency for bottom-up attentional control," suggesting that "they may be sacrificing performance on the primary task to let in other sources of information." Intensive multitaskers are "suckers for irrelevancy," commented Clifford Nass, the Stanford professor who led the research. "Everything distracts them."[53] Michael Merzenich offers an even bleaker assessment. As we multitask online, he says, we are "training our brains to pay attention to the crap." The consequences for our intellectual lives may prove "deadly."[54]

The mental functions that are losing the "survival of the busiest" brain cell battle are those that support calm, linear thought—the ones we use in traversing a lengthy narrative or an involved argument, the ones we draw on when we reflect on our experiences or contemplate an outward or inward phenomenon. The winners are those functions that help us speedily locate, categorize, and assess disparate bits of information in a variety of forms, that let us maintain our mental bearings while being bombarded by stimuli. These functions are, not coincidentally, very similar to the ones performed by computers, which are programmed for the high-speed transfer of data in and out of memory. Once again, we seem to be taking on the characteristics of a popular new intellectual technology.

ON THE EVENING of April 18, 1775, Samuel Johnson accompanied his friends James Boswell and Joshua Reynolds on a visit to Richard Owen Cambridge's grand villa on the banks of the Thames outside London. They were shown into the library, where Cambridge was waiting to meet them, and after a brief greeting Johnson darted to the shelves and began silently reading the spines of the volumes arrayed there. "Dr. Johnson," said Cambridge, "it seems odd that one should have such a desire to look at the backs of books." Johnson, Boswell would later recall, "instantly started from his reverie, wheeled about, and replied, 'Sir, the reason is very plain. Knowledge is of two kinds. We know a subject ourselves, or we know where we can find information upon it.'"[55]

The Net grants us instant access to a library of information unprecedented in its size and scope, and it makes it easy for us to sort through that library—to find, if not exactly what we were looking for, at least something sufficient for our immediate purposes. What the Net diminishes is Johnson's primary kind of knowledge: the ability to know, in depth, a subject for ourselves, to construct within our own minds the rich and idiosyncratic set of connections that give rise to a singular intelligence.

a digression

---

on the buoyancy of IQ scores

THIRTY YEARS AGO, James Flynn, then the head of the political science department at New Zealand's University of Otago, began studying historical records of IQ tests. As he dug through the numbers, stripping out the various scoring adjustments that had been made through the years, he discovered something startling: IQ scores had been rising steadily—and pretty much everywhere—throughout the century. Controversial when originally reported, the Flynn effect, as the phenomenon came to be called, has been confirmed by many subsequent studies. It's real.

Ever since Flynn made his discovery, it has provided a ready-made brickbat to hurl at anyone who suggests that our intellectual powers may be on the wane: *If we're so dumb, why do we keep getting smarter?* The Flynn effect has been used to defend TV shows, video games, personal computers, and, most recently, the Internet. Don Tapscott, in *Grown Up Digital*, his paean to the first generation of "digital natives," counters arguments that the extensive use of digital media may be dumbing kids down by pointing out, with a nod to Flynn, that "raw IQ scores have been going up three points a decade since World War II."[1]

Tapscott's right about the numbers, and we should certainly be heartened by the rise in IQ scores, particularly since the gains have been sharpest among segments of the population whose scores have lagged in the past. But there are good reasons to be skeptical of any claim that the Flynn effect proves that people are "smarter" today than they used to be or that the Internet is boosting the general intelligence of the human race. For one thing, as Tapscott himself notes, IQ scores have been going up for a very long time—since well before World War II, in fact—and the pace of increase has remained remarkably stable, varying only slightly from decade to decade. That pattern suggests that the rise probably reflects a deep and persistent change in some aspect of society rather than any particular recent event or technology. The fact that the Internet began to come into widespread use only about ten years ago makes it all the more unlikely that it has been a significant force propelling IQ scores upward.

Other measures of intelligence don't show anything like the gains we've seen in overall IQ scores. In fact, even IQ tests have been sending mixed signals. The tests have different sections, which measure different aspects of intelligence, and performance on them has varied widely. Most of the increase in overall scores can be attributed to strengthening performance in tests involving the mental rotation of geometric forms, the identification of similarities between disparate objects, and the arrangement of shapes into logical sequences. Tests of memorization, vocabulary, general knowledge, and even basic arithmetic have shown little or no improvement.

Scores on other common tests designed to measure intellectual skills also seem to be either stagnant or declining. Scores on PSAT exams, which are given to high school juniors throughout the United States, did not increase at all during the years from 1999 to 2008, a time when Net use in homes and schools was expanding dramatically. In fact, while the average math scores held fairly steady during that period, dropping a fraction of a point, from 49.2 to 48.8, scores on the verbal portions of the test declined significantly. The average critical-reading score fell 3.3 percent, from 48.3 to 46.7, and the average writing-skills score dropped an even steeper 6.9 percent, from 49.2 to 45.8.[2] Scores on the verbal sections of the SAT tests given to college-bound students have also been dropping. A 2007 report from the U.S. Department of Education showed that twelfth-graders' scores on tests of three different kinds of reading—for performing a task, for gathering information, and for literary experience—fell between 1992 and 2005. Literary reading aptitude suffered the largest decline, dropping twelve percent.[3]

There are signs, as well, that the Flynn effect may be starting to fade even as Web use picks up. Research in Norway and Denmark shows that the rise in intelligence test scores began to slow in those countries during the 1970s and '80s and that since the mid-1990s scores have either remained steady or fallen slightly.[4] In the United Kingdom, a 2009 study revealed that the IQ scores of teenagers dropped by two points between 1980 and 2008, after decades of gains.[5] Scandinavians and Britons have been among the world's pacesetters in adopting high-speed Internet service and using multipurpose mobile phones. If digital media were boosting IQ scores, you'd expect to see particularly strong evidence in their results.

So what is behind the Flynn effect? Many theories have been offered, from smaller families to better nutrition to the expansion of formal education, but the explanation that seems most credible comes from James Flynn himself. Early in his research, he realized that his findings presented a couple of paradoxes. First, the steepness of the rise in test scores during the twentieth century suggests that our forebears must have been dimwits, even though everything we know about them tells us otherwise. As Flynn wrote in his book *What Is Intelligence?*, "If IQ gains are in any sense real, we are driven to the absurd conclusion that a majority of our ancestors were mentally retarded."[6] The second paradox stems from the disparities in the scores on different sections of IQ tests: "How can people get more intelligent and have no larger vocabularies, no larger stores of general information, no greater ability to solve arithmetical problems?"[7]

After mulling over the paradoxes for many years, Flynn came to the conclusion that the gains in IQ scores have less to do with an increase in general intelligence than with a transformation in the way people think about intelligence. Up until the end of the nineteenth century, the scientific view of intelligence, with its stress on classification, correlation, and abstract reasoning, remained fairly rare, limited to those who attended or taught at universities. Most people continued to see intelligence as a matter of deciphering the workings of nature and solving practical problems—on the farm, in the factory, at home. Living in a world of substance rather than symbol, they had little cause or opportunity to think about abstract shapes and theoretical classification schemes.

But, Flynn realized, that all changed over the course of the last century when, for economic, technological, and educational reasons, abstract reasoning moved into the mainstream. Everyone began to wear, as Flynn colorfully puts it, the same "scientific spectacles" that were worn by the original developers of IQ tests.[8] Once he had that insight, Flynn recalled in a 2007 interview, "I began to feel that I was bridging the gulf between our minds and the minds of our ancestors. We weren't more intelligent than they, but we had learnt to apply our intelligence to a new set of problems. We had detached logic from the concrete, we were willing to deal with the hypothetical, and we thought the world was a place to be classified and understood scientifically rather than to be manipulated."[9]

Patricia Greenfield, the UCLA psychologist, came to a similar conclusion in her *Science* article on media and intelligence. Noting that the rise in IQ scores "is concentrated in nonverbal IQ performance," which is "mainly tested through visual tests," she attributed the Flynn effect to an array of factors, from urbanization to the growth in "societal complexity," all of which "are part and parcel of the worldwide movement from smaller-scale, low-tech communities with subsistence economies toward large-scale, high-tech societies with commercial economies."[10]

We're not smarter than our parents or our parents' parents. We're just smart in different ways. And that influences not only how we see the world but also how we raise and educate our children. This social revolution in how we think about thinking explains why we've become ever more adept at working out the problems in the more abstract and visual sections of IQ tests while making little or no progress in expanding our personal knowledge, bolstering our basic academic skills, or improving our ability to communicate compli- cated ideas clearly. We're trained, from infancy, to put things into categories, to solve puzzles, to think in terms of symbols in space. Our use of personal computers and the Internet may well be rein- forcing some of those mental skills and the corresponding neural circuits by strengthening our visual acuity, particularly our ability to speedily evaluate objects and other stimuli as they appear in the abstract realm of a computer screen. But, as Flynn stresses, that doesn't mean we have "better brains." It just means we have differ- ent brains.[11]

Eight

———

THE CHURCH OF GOOGLE

Not long after Nietzsche bought his mechanical writing ball, an earnest young man named Frederick Winslow Taylor carried a stopwatch into the Midvale Steel plant in Philadelphia and began a historic series of experiments aimed at boosting the efficiency of the plant's machinists. With the grudging approval of Midvale's owners, Taylor recruited a group of factory hands, set them to work on various metalworking machines, and recorded and timed their every movement. By breaking down each job into a sequence of small steps and then testing different ways of performing them, he created a set of precise instructions—an "algorithm," we might say today—for how each worker should work. Midvale's employees grumbled about the strict new regime, claiming that it turned them into little more than automatons, but the factory's productivity soared.[1]

More than a century after the invention of the steam engine, the Industrial Revolution had at last found its philosophy and its philosopher. Taylor's tight industrial choreography—his "system," as he liked to call it—was embraced by manufacturers throughout the country and, in time, around the world. Seeking maximum speed, maximum efficiency, and maximum output, factory owners used time-and-motion studies to organize their work and configure the jobs of their workers. The goal, as Taylor defined it in his celebrated 1911 treatise *The Principles of Scientific Management*, was to identify and adopt, for every job, the "one best method" of work and thereby to effect "the gradual substitution of science for rule of thumb throughout the mechanic arts."[2] Once his system was applied to all acts of manual labor, Taylor assured his many followers, it would bring about a restructuring not only of industry but of society, creating a utopia of perfect efficiency. "In the past the man has been first," he declared; "in the future the system must be first."[3]

Taylor's system of measurement and optimization is still very much with us; it remains one of the underpinnings of industrial manufacturing. And now, thanks to the growing power that computer engineers and software coders wield over our intellectual and social lives, Taylor's ethic is beginning to govern the realm of the mind as well. The Internet is a machine designed for the efficient, automated collection, transmission, and manipulation of information, and its legions of programmers are intent on finding the "one best way"—the perfect algorithm—to carry out the mental movements of what we've come to describe as knowledge work.

Google's Silicon Valley headquarters—the Googleplex—is the Internet's high church, and the religion practiced inside its walls is Taylorism. The company, says CEO Eric Schmidt, is "founded around the science of measurement." It is striving to "systematize everything" it does.[4] "We try to be very data-driven, and quantify everything," adds another Google executive, Marissa Mayer. "We live in a world of numbers."[5] Drawing on the terabytes of behavioral data it collects through its search engine and other sites, the company carries out thousands of experiments a day and uses the results to refine the algorithms that increasingly guide how all of us find information and extract meaning from it.[6] What Taylor did for the work of the hand, Google is doing for the work of the mind.

The company's reliance on testing is legendary. Although the

design of its Web pages may appear simple, even austere, each element has been subjected to exhaustive statistical and psychological
research. Using a technique called "split A/B testing," Google continually introduces tiny permutations in the way its sites look and
operate, shows different permutations to different sets of users, and
then compares how the variations influence the users' behavior—
how long they stay on a page, the way they move their cursor about
the screen, what they click on, what they don't click on, where they
go next. In addition to the automated online tests, Google recruits
volunteers for eye-tracking and other psychological studies at its
in-house "usability lab." Because Web surfers evaluate the contents
of pages "so quickly that they make most of their decisions unconsciously," remarked two Google researchers in a 2009 blog post about
the lab, monitoring their eye movements "is the next best thing to
actually being able to read their minds."[7] Irene Au, the company's
director of user experience, says that Google relies on "cognitive
psychology research" to further its goal of "making people use their
computers more efficiently."[8]

Subjective judgments, including aesthetic ones, don't enter into
Google's calculations. "On the web," says Mayer, "design has become
much more of a science than an art. Because you can iterate so
quickly, because you can measure so precisely, you can actually find
small differences and mathematically learn which one is right."[9] In
one famous trial, the company tested forty-one different shades of
blue on its toolbar to see which shade drew the most clicks from visitors. It carries out similarly rigorous experiments on the text it puts
on its pages. "You have to try and make words less human and more
a piece of the machinery," explains Mayer.[10]

In his 1993 book *Technopoly*, Neil Postman distilled the main
tenets of Taylor's system of scientific management. Taylorism, he
wrote, is founded on six assumptions: "that the primary, if not the
only, goal of human labor and thought is efficiency; that technical
calculation is in all respects superior to human judgment; that in
fact human judgment cannot be trusted, because it is plagued by lax-

ity, ambiguity, and unnecessary complexity; that subjectivity is an obstacle to clear thinking; that what cannot be measured either does not exist or is of no value; and that the affairs of citizens are best guided and conducted by experts."[11] What's remarkable is how well Postman's summary encapsulates Google's own intellectual ethic. Only one tweak is required to bring it up to date. Google doesn't believe that the affairs of citizens are best guided by experts. It believes that those affairs are best guided by software algorithms— which is exactly what Taylor would have believed had powerful digital computers been around in his day.

Google also resembles Taylor in the sense of righteousness it brings to its work. It has a deep, even messianic faith in its cause. Google, says its CEO, is more than a mere business; it is a "moral force."[12] The company's much-publicized "mission" is "to organize the world's information and make it universally accessible and useful."[13] Fulfilling that mission, Schmidt told the *Wall Street Journal* in 2005, "will take, current estimate, 300 years."[14] The company's more immediate goal is to create "the perfect search engine," which it defines as "something that understands exactly what you mean and gives you back exactly what you want."[15] In Google's view, information is a kind of commodity, a utilitarian resource that can, and should, be mined and processed with industrial efficiency. The more pieces of information we can "access" and the faster we can distill their gist, the more productive we become as thinkers. Anything that stands in the way of the speedy collection, dissection, and transmission of data is a threat not only to Google's business but to the new utopia of cognitive efficiency it aims to construct on the Internet.

GOOGLE WAS BORN of an analogy—Larry Page's analogy. The son of one of the pioneers of artificial intelligence, Page was surrounded by computers from an early age—he recalls being "the first kid in my elementary school to turn in a word-processed document"[16]—and went on to study engineering as an undergraduate at the University of Michigan. His friends remember him as being ambitious, smart, and "nearly obsessed with efficiency."[17] While serving as president of Michigan's engineering honor society, he spearheaded a brash, if ultimately futile, campaign to convince the school's administrators to build a monorail through the campus. In the fall of 1995, Page headed to California to take a prized spot in Stanford University's doctoral program in computer science. Even as a young boy, he had dreamed of creating a momentous invention, something that "would change the world."[18] He knew there was no better place than Stanford, Silicon Valley's frontal cortex, to make the dream come true.

It took only a few months for Page to land on a topic for his dissertation: the vast new computer network called the World Wide Web. Launched on the Internet just four years earlier, the Web was growing explosively—it had half a million sites and was adding more than a hundred thousand new ones every month—and the network's incredibly complex and ever-shifting arrangement of nodes and links had come to fascinate mathematicians and computer scientists. Page had an idea that he thought might unlock some of its secrets. He had realized that the links on Web pages are analogous to the citations in academic papers. Both are signifiers of value. When a scholar, in writing an article, makes a reference to a paper published by another scholar, she is vouching for the importance of that other paper. The more citations a paper garners, the more prestige it gains in its field. In the same way, when a person with a Web page links to someone else's page, she is saying that she thinks the other page is important. The value of any Web page, Page saw, could be gauged by the links coming into it.

Page had another insight, again drawing on the citations analogy: not all links are created equal. The authority of any Web page can be gauged by how many incoming links it attracts. A page with a lot of incoming links has more authority than a page with only one or two. The greater the authority of a Web page, the greater the worth of its own outgoing links. The same is true in academia: earning a citation from a paper that has itself been much cited is more valuable than receiving one from a less cited paper. Page's analogy led him to realize that the relative value of any Web page could be estimated through a mathematical analysis of two factors: the number of incoming links the page attracted and the authority of the sites that were the sources of those links. If you could create a database of all the links on the Web, you would have the raw material to feed into a software algorithm that could evaluate and rank the value of all the pages on the Web. You would also have the makings of the world's most powerful search engine.

The dissertation never got written. Page recruited another Stanford graduate student, a math prodigy named Sergey Brin who had a deep interest in data mining, to help him build his search engine. In the summer of 1996, an early version of Google—then called Back-Rub—debuted on Stanford's Web site. Within a year, BackRub's traffic had overwhelmed the university's network. If they were going to turn their search service into a real business, Page and Brin saw, they were going to need a lot of money to buy computing gear and network bandwidth. In the summer of 1998, a wealthy Silicon Valley investor came to the rescue, cutting them a check for a hundred grand. They moved their budding company out of their dorms and into a couple of spare rooms in a friend-of-a-friend's house in nearby Menlo Park. In September they incorporated as Google Inc. They chose the name—a play on *googol*, the word for the number ten raised to the hundredth power—to highlight their goal of organizing "a seemingly infinite amount of information on the web." In December, an article in *PC Magazine* praised the new search engine with the quirky name, saying it "has an uncanny knack for returning extremely relevant results."[19]

Thanks to that knack, Google was soon processing most of the millions—and then billions—of Internet searches being conducted every day. The company became fabulously successful, at least as measured by the traffic running through its site. But it faced the same problem that had doomed many dot-coms: it hadn't been able to figure out how to turn a profit from all that traffic. No one would pay to search the Web, and Page and Brin were averse to injecting advertisements into their search results, fearing it would corrupt Google's pristine mathematical objectivity. "We expect," they had written in a scholarly paper early in 1998, "that advertising-funded search engines will be inherently biased towards the advertisers and away from the needs of the consumers."[20]

But the young entrepreneurs knew that they would not be able to live off the largesse of venture capitalists forever. Late in 2000, they came up with a clever plan for running small, textual advertisements alongside their search results—a plan that would require only a modest compromise of their ideals. Rather than selling advertising space for a set price, they decided to auction the space off. It wasn't an original idea—another search engine, GoTo, was already auctioning ads—but Google gave it a new spin. Whereas GoTo ranked its search ads according to the size of advertisers' bids—the higher the bid, the more prominent the ad—Google in 2002 added a second criterion. An ad's placement would be determined not only by the amount of the bid but by the frequency with which people actually clicked on the ad. That innovation ensured that Google's ads would remain, as the company put it, "relevant" to the topics of searches. Junk ads would automatically be screened from the system. If searchers didn't find an ad relevant, they wouldn't click on it, and it would eventually disappear from Google's site.

The auction system, named AdWords, had another, very important result: by tying ad placement to clicks, it increased click-through rates substantially. The more often people clicked on an ad, the more frequently and prominently the ad would appear on search result pages, bringing even more clicks. Since advertisers paid Google by the click, the company's revenues soared. The AdWords system proved so lucrative that many other Web publishers contracted with Google to place its "contextual ads" on their sites as well, tailoring the ads to the content of each page. By the end of the decade, Google was not just the largest Internet company in the world; it was one of the largest media companies, taking in more than $22 billion in

sales a year, almost all of it from advertising, and turning a profit
of about $8 billion. Page and Brin were each worth, on paper, more
than $10 billion.

Google's innovations have paid off for its founders and investors.
But the biggest beneficiaries have been Web users. Google has succeeded in making the Internet a far more efficient informational
medium. Earlier search engines tended to get clogged with data as
the Web expanded—they couldn't index the new content, much less
separate the wheat from the chaff. Google's engine, by contrast, has
been engineered to produce better results as the Web grows. The
more sites and links Google evaluates, the more precisely it can classify pages and rank their quality. And as traffic increases, Google is
able to collect more behavioral data, allowing it to tailor its search
results and advertisements ever more precisely to users' needs and
desires. The company has also invested many billions of dollars in
building computer-packed data centers around the world, ensuring
that it can deliver search results to its users in milliseconds. Google's
popularity and profitability are well deserved. The company plays an
invaluable role in helping people navigate the hundreds of billions of
pages that now populate the Web. Without its search engine, and the
other engines that have been built on its model, the Internet would
have long ago become a Tower of Digital Babel.

But Google, as the supplier of the Web's principal navigational
tools, also shapes our relationship with the content that it serves
up so efficiently and in such profusion. The intellectual technologies it has pioneered promote the speedy, superficial skimming of
information and discourage any deep, prolonged engagement with a
single argument, idea, or narrative. "Our goal," says Irene Au, "is to
get users in and out really quickly. All our design decisions are based
on that strategy."[21] Google's profits are tied directly to the velocity
of people's information intake. The faster we surf across the surface
of the Web—the more links we click and pages we view—the more
opportunities Google gains to collect information about us and to
feed us advertisements. Its advertising system, moreover, is explic-

itly designed to figure out which messages are most likely to grab our attention and then to place those messages in our field of view. Every click we make on the Web marks a break in our concentration, a bottom-up disruption of our attention—and it's in Google's economic interest to make sure we click as often as possible. The last thing the company wants is to encourage leisurely reading or slow, concentrated thought. Google is, quite literally, in the business of distraction.

GOOGLE MAY YET turn out to be a flash in the pan. The lives of Internet companies are rarely nasty or brutish, but they do tend to be short. Because their businesses are ethereal, constructed of invisible strands of software code, their defenses are fragile. All it takes to render a thriving online business obsolete is a sharp programmer with a fresh idea. The invention of a more precise search engine or a better way to circulate ads through the Net could spell ruin for Google. But no matter how long the company is able to maintain its dominance over the flow of digital information, its intellectual ethic will remain the general ethic of the Internet as a medium. Web publishers and toolmakers will continue to attract traffic and make money by encouraging and feeding our hunger for small, rapidly dispensed pieces of information.

The history of the Web suggests that the velocity of data will only increase. During the 1990s, most online information was found on so-called static pages. They didn't look all that different from the pages in magazines, and their content remained relatively fixed. The trend since then has been to make pages ever more "dynamic," updating them regularly and often automatically with new content. Specialized blogging software, introduced in 1999, made rapid-fire publishing simple for everyone, and the most successful bloggers soon found that they needed to post many items a day to keep fickle readers engaged. News sites followed suit, serving up fresh stories around the clock. RSS readers, which became popular around 2005, allowed sites to "push" headlines and other bits of information to Web users, putting an even greater premium on the frequency of information delivery.

The greatest acceleration has come recently, with the rise of social networks like LinkedIn, Facebook, and Twitter. These companies are dedicated to providing their millions of members with a never-ending "stream" of "real-time updates," brief messages about, as a Twitter slogan puts it, "what's happening *right now*." By turning intimate messages—once the realm of the letter, the phone call, the whisper—into fodder for a new form of mass media, the social networks have given people a compelling new way to socialize and stay in touch. They've also placed a whole new emphasis on immediacy. A "status update" from a friend, co-worker, or favorite celebrity loses its currency within moments of being issued. To be up to date requires the continual monitoring of message alerts. The competition among the social networks to deliver ever-fresher and more plentiful messages is fierce. When, in early 2009, Facebook responded to Twitter's rapid growth by announcing that it was revamping its site to, as it put it, "increase the pace of the stream," its founder and chief executive, Mark Zuckerberg, assured its quarter of a billion members that the company would "continue making the flow of information even faster."[22] Unlike early book printers, who had strong economic incentives to promote the reading of older works as well as recent ones, online publishers battle to distribute the newest of the new.

Google hasn't been sitting still. To combat the upstarts, it has been revamping its search engine to ratchet up its speed. The quality of a page, as determined by the links coming into it, is no longer Google's chief criterion in ranking search results. In fact, it's now only one of two hundred different "signals" that the company monitors and measures, according to Amit Singhal, a top Google engineer.[23] One of its major recent thrusts has been to place a greater priority on what it calls the "freshness" of the pages it recommends. Google not only identifies new or revised Web pages much more quickly than it used to—it now checks the most popular sites for updates every few seconds rather than every few days—but for many searches it skews its results to favor newer pages over older ones. In May 2009, the company introduced a new twist to its search service, allowing users to bypass considerations of quality entirely and have results ranked according to how recently the information was posted to the Web. A few months later, it announced a "next-generation architecture" for its search engine that bore the telling code name Caffeine.[24] Citing Twitter's achievements in speeding the flow of data, Larry Page said that Google wouldn't be satisfied until it is able "to index the Web every second to allow real-time search."[25]

The company is also striving to further expand its hold on Web users and their data. With the billions in profits churned out by AdWords, it has been able to diversify well beyond its original focus on searching Web pages. It now has specialized search services for, among other things, images, videos, news stories, maps, blogs, and academic journals, all of which feed into the results supplied by its main search engine. It also offers computer operating systems, such as Android for smartphones and Chrome for PCs, as well as a slew of online software programs, or "apps," including e-mail, word processing, blogging, photo storage, feed reading, spreadsheets, calendars, and Web hosting. Google Wave, an ambitious social-networking service launched at the end of 2009, allows people to monitor and update various multimedia message threads on a single densely packed page, which refreshes its contents automatically and almost instantaneously. Wave, says one reporter, "turns conversations into fast-moving group streams-of-consciousness."[26]

The company's seemingly boundless expansiveness has been a matter of much discussion, particularly among management scholars and business reporters. The breadth of its influence and activity is often interpreted as evidence that it is an entirely new species of business, one that transcends and redefines all traditional categories. But while Google is an unusual company in many ways, its business strategy is not quite as mysterious as it seems. Google's protean appearance is not a reflection of its main business: selling and distributing online ads. Rather, it stems from the vast number of "complements" to that business. Complements are, in economic terms, any products or services that tend be purchased or consumed together, such as hot dogs and mustard or lamps and lightbulbs. For Google, everything that happens on the Internet is a complement to its main business. As people spend more time and do more things online, they see more ads and they disclose more information about themselves—and Google rakes in more money. As additional products and services have come to be delivered digitally over computer networks—entertainment, news, software applications, financial transactions, phone calls—Google's range of complements has extended into ever more industries.

Because the sales of complementary products rise in tandem, a company has a strong strategic interest in reducing the cost and expanding the availability of the complements to its main product. It's not too much of an exaggeration to say that a company would like all complements to be given away. If hot dogs were free, mustard sales would skyrocket. It's this natural drive to reduce the cost of complements that, more than anything else, explains Google's business strategy. Nearly everything the company does is aimed at reducing the cost and expanding the scope of Internet use. Google wants information to be free because, as the cost of information falls, we all spend more time looking at computer screens and the company's profits go up.

Most of Google's services are not profitable in themselves. Industry analysts estimate, for example, that YouTube, which Google bought for $1.65 billion in 2006, lost between $200 million and $500 million in 2009.[27] But because popular services like YouTube enable Google to collect more information, to funnel more users toward its search engine, and to prevent would-be competitors from gaining footholds in its markets, the company is able to justify the cost of launching them. Google has let it be known that it won't be satisfied until it stores "100% of user data."[28] Its expansionary zeal isn't just about money, though. The steady colonization of additional types of content also furthers the company's mission of making the world's information "universally accessible and useful." Its ideals and its business interests converge in one overarching goal: to digitize ever more types of information, move the information onto the Web, feed it into its database, run it through its classification and ranking algorithms, and dispense it in what it calls "snippets" to Web surfers, preferably with ads in tow. With each expansion of Google's ambit, its Taylorist ethic gains a tighter hold on our intellectual lives.

THE MOST AMBITIOUS of Google's initiatives—what Marissa Mayer calls its "moon shot"[29]—is its effort to digitize all the books ever printed and make their text "discoverable and searchable online."[30] The program began in secret in 2002, when Larry Page set up a digital scanner in his office in the Googleplex and, to the beat of a metronome, spent a half hour methodically scanning the pages of a three-hundred-page book. He wanted to get a rough sense of how long it would take "to digitally scan every book in the world." The next year, a Google employee was sent to Phoenix to buy a pile of old books at a charity sale. Once carted back to the Googleplex, the volumes became the test subjects in a series of experiments that led to the development of a new "high-speed" and "non-destructive" scanning technique. The ingenious system, which involves the use of stereoscopic infrared cameras, is able to automatically correct for the bowing of pages that occurs when a book is opened, eliminating any distortion of the text in the scanned image.[31] At the same time, a team of Google software engineers was fine-tuning a sophisticated character recognition program able to handle "odd type sizes, unusual fonts or other unexpected peculiarities—in 430 different languages." Another group of Google employees spread out to visit leading libraries and book publishers to gauge their interest in having Google digitize their books.[32]

In the fall of 2004, Page and Brin formally announced the Google Print program (it would later be renamed Google Book Search) at

the Frankfurt Book Fair, an event that since Gutenberg's day has
been the publishing industry's chief annual gathering. More than
a dozen trade and academic presses signed on as Google's partners,
including such top names as Houghton Mifflin, McGraw-Hill, and
the university presses of Oxford, Cambridge, and Princeton. Five of
the world's most prestigious libraries, including Harvard's Widener,
Oxford's Bodleian, and the New York Public Library, also agreed to
collaborate in the effort. They granted Google permission to begin
scanning the contents of their stacks. By the end of the year, the
company already had the text of an estimated hundred thousand
books in its data bank.

Not everyone was happy with the library scanning project. Google
was not just scanning old books that had fallen out of copyright pro-
tection. It was also scanning newer books that, while often out of
print, were still the copyrighted property of their authors or publishers. Google made it clear that it had no intention of tracking
down and securing the consent of the copyright holders in advance.
Rather, it would proceed to scan all the books and include them
in its database unless a copyright owner sent it a formal written
request to exclude a particular book. On September 20, 2005, the
Authors Guild, along with three prominent writers acting individu-
ally, sued Google, alleging that the scanning program entailed "massive copyright infringement."[33] A few weeks later, the Association
of American Publishers filed another lawsuit against the company,
demanding that it stop scanning the libraries' collections. Google
fired back, launching a public relations offensive to publicize the
societal benefits of Google Book Search. In October, Eric Schmidt
wrote an op-ed column for the *Wall Street Journal* that portrayed the
book digitization effort in terms at once stirring and vainglorious:
"Imagine the cultural impact of putting tens of millions of previously inaccessible volumes into one vast index, every word of which
is searchable by anyone, rich and poor, urban and rural, First World
and Third, *en toute langue*—and all, of course, entirely for free."[34]

The suits proceeded. After three years of negotiations, during which Google scanned some seven million additional books, six million of which were still under copyright, the parties reached a settlement. Under the terms of the accord, announced in October 2008, Google agreed to pay $125 million to compensate the owners of the copyrights in the works that it had already scanned. It also agreed to set up a payment system that would give authors and publishers a cut of advertising and other revenues earned from the Google Book Search service in the years ahead. In return for the concessions, the authors and publishers gave Google their okay to proceed with its plan to digitize all the world's books. The company would also be "authorized to, in the United States, sell subscriptions to [an] Institutional Subscription Database, sell individual Books, place advertisements on Online Book Pages, and make other commercial uses of Books."[35]

The proposed settlement set off another, even fiercer controversy. The terms appeared to give Google a monopoly over the digital versions of millions of so-called orphan books—those whose copyright owners are unknown or can't be found. Many libraries and schools feared that, without competition, Google would be able to raise the subscription fees for its book database as high as it liked. The American Library Association, in a court filing, warned that the company might "set the price of the subscription at a profit-maximizing point beyond the reach of many libraries."[36] The U.S. Justice Department and Copyright Office both criticized the deal, contending it would give Google too much power over the future market for digital books.

Other critics had a related but more general worry: that commercial control over the distribution of digital information would inevitably lead to restrictions on the flow of knowledge. They were suspicious of Google's motives, despite its altruistic rhetoric. "When businesses like Google look at libraries, they do not merely see temples of learning," wrote Robert Darnton, who, in addition to teaching at Harvard, oversees its library system. "They see potential assets or what they call 'content,' ready to be mined." Although Google "has pursued a laudable goal" in "promoting access to information," con-

ceded Darnton, granting a profit-making enterprise a monopoly "not
of railroads or steel but of access to information" would entail too
great a risk. "What will happen if its current leaders sell the company
or retire?" he asked. "What will happen if Google favors profitability
over access?"[37] By the end of 2009, the original agreement had been
abandoned, and Google and the other parties were trying to win support for a slightly less sweeping alternative.

The debate over Google Book Search is illuminating for several
reasons. It reveals how far we still have to go to adapt the spirit and
letter of copyright law, particularly its fair-use provisions, to the digital age. (The fact that some of the publishing firms that were parties to the lawsuit against Google are also partners in Google Book
Search testifies to the murkiness of the current situation.) It also
tells us much about Google's high-flown ideals and the high-handed
methods it sometimes uses to pursue them. One observer, the law-
yer and technology writer Richard Koman, argued that Google "has
become a true believer in its own goodness, a belief which justi-
fies its own set of rules regarding corporate ethics, anti-competition,
customer service and its place in society."[38]

Most important of all, the controversy makes clear that the
world's books *will* be digitized—and that the effort is likely to proceed quickly. The argument about Google Book Search has nothing
to do with the wisdom of scanning printed books into a database; it
has to do with the control and commercialization of that database.
Whether or not Google ends up being the sole proprietor of what
Darnton calls "the largest library in the world," that library is going
to be constructed; and its digital volumes, fed through the Net into
every library on earth, will in time supplant many of the physical
books that have long been stored on shelves.[39] The practical benefits
of making books "discoverable and searchable online" are so great
that it's hard to imagine anyone opposing the effort. The digitization of old books, as well as ancient scrolls and other documents,
is already opening exciting new avenues for research into the past.

Some foresee "a second Renaissance" of historical discovery.[40] As Darnton says, "Digitize we must."

But the inevitability of turning the pages of books into online images should not prevent us from considering the side effects. To make a book discoverable and searchable online is also to dismember it. The cohesion of its text, the linearity of its argument or narrative as it flows through scores of pages, is sacrificed. What that ancient Roman craftsman wove together when he created the first codex is unstitched. The quiet that was "part of the meaning" of the codex is sacrificed as well. Surrounding every page or snippet of text on Google Book Search is a welter of links, tools, tabs, and ads, each eagerly angling for a share of the reader's fragmented attention.

For Google, with its faith in efficiency as the ultimate good and its attendant desire "to get users in and out really quickly," the unbinding of the book entails no loss, only gain. Google Book Search manager Adam Mathes grants that "books often live a vibrant life offline," but he says that they'll be able to "live an even more exciting life online."[41] What does it mean for a book to lead a more exciting life? Searchability is only the beginning. Google wants us, it says, to be able to "slice and dice" the contents of the digitized books we discover, to do all the "linking, sharing, and aggregating" that are routine with Web content but that "you can't easily do with physical books." The company has already introduced a cut-and-paste tool that "lets you easily clip and publish passages from public domain books on your blog or website."[42] It has also launched a service it calls Popular Passages, which highlights brief excerpts from books that have been quoted frequently, and for some volumes it has begun displaying "word clouds" that allow a reader to, as the company says, "explore a book in 10 seconds."[43] It would be silly to complain about such tools. They *are* useful. But they also make clear that, for Google, the real value of a book is not as a self-contained literary work but as another pile of data to be mined. The great library that Google is rushing to create shouldn't be confused with the libraries we've known up until now. It's not a library of books. It's a library of snippets.

The irony in Google's effort to bring greater efficiency to reading is that it undermines the very different kind of efficiency that the technology of the book brought to reading—and to our minds—in the first place. By freeing us from the struggle of decoding text, the form that writing came to take on a page of parchment or paper enabled us to become deep readers, to turn our attention, and our brain power, to the interpretation of meaning. With writing on the screen, we're still able to decode text quickly—we read, if anything, faster than ever—but we're no longer guided toward a deep, personally constructed understanding of the text's connotations. Instead, we're hurried off toward another bit of related information, and then another, and another. The strip-mining of "relevant content" replaces the slow excavation of meaning.

IT WAS A warm summer morning in Concord, Massachusetts. The year was 1844. An aspiring novelist named Nathaniel Hawthorne was sitting in a small clearing in the woods, a particularly peaceful spot known around town as Sleepy Hollow. Deep in concentration, he was attending to every passing impression, turning himself into what Emerson, the leader of Concord's Transcendentalist movement, had eight years earlier termed a "transparent eyeball." Hawthorne saw, as he would record in his notebook later that day, how "sunshine glimmers through shadow, and shadow effaces sunshine, imaging that pleasant mood of mind where gayety and pensiveness intermingle." He felt a slight breeze, "the gentlest sigh imaginable, yet with a spiritual potency, insomuch that it seems to penetrate, with its mild, ethereal coolness, through the outward clay, and breathe upon the spirit itself, which shivers with gentle delight." He smelled on the breeze a hint of "the fragrance of the white pines." He heard "the striking of the village clock" and "at a distance mowers whetting their scythes," though "these sounds of labor, when at a proper remoteness, do but increase the quiet of one who lies at his ease, all in a mist of his own musings."

Abruptly, his reverie was broken:

> But, hark! there is the whistle of the locomotive,—the long shriek, harsh above all other harshness, for the space of a mile cannot mollify it into harmony. It tells a story of busy men, citizens from the hot street, who have come to spend a day in a country village,—men of business,—in short, of all unquietness; and no wonder that it gives such a startling shriek, since it brings the noisy world into the midst of our slumbrous peace.[44]

Leo Marx opens *The Machine in the Garden*, his classic 1964 study of technology's influence on American culture, with a recounting of Hawthorne's morning in Sleepy Hollow. The writer's real subject, Marx argues, is "the landscape of the psyche" and in particular "the contrast between two conditions of consciousness." The quiet clearing in the woods provides the solitary thinker with "a singular insulation from disturbance," a protected space for reflection. The clamorous arrival of the train, with its load of "busy men," brings "the psychic dissonance associated with the onset of industrialism."[45] The contemplative mind is overwhelmed by the noisy world's mechanical busyness.

The stress that Google and other Internet companies place on the efficiency of information exchange as the key to intellectual progress is nothing new. It's been, at least since the start of the Industrial Revolution, a common theme in the history of the mind. It provides a strong and continuing counterpoint to the very different view, promulgated by the American Transcendentalists as well as the earlier English Romantics, that true enlightenment comes only through contemplation and introspection. The tension between the two perspectives is one manifestation of the broader conflict between, in Marx's terms, "the machine" and "the garden"—the industrial ideal and the pastoral ideal—that has played such an important role in shaping modern society.

When carried into the realm of the intellect, the industrial ideal
of efficiency poses, as Hawthorne understood, a potentially mortal
threat to the pastoral ideal of meditative thought. That doesn't mean
that promoting the rapid discovery and retrieval of information is
bad. It's not. The development of a well-rounded mind requires both
an ability to find and quickly parse a wide range of information and
a capacity for open-ended reflection. There needs to be time for efficient data collection and time for inefficient contemplation, time
to operate the machine and time to sit idly in the garden. We need to
work in Google's "world of numbers," but we also need to be able to
retreat to Sleepy Hollow. The problem today is that we're losing our
ability to strike a balance between those two very different states of
mind. Mentally, we're in perpetual locomotion.

Even as Gutenberg's press was making the literary mind the general mind, it was setting in motion the process that now threatens
to render the literary mind obsolete. When books and periodicals
began to flood the marketplace, people for the first time felt overwhelmed by information. Robert Burton, in his 1628 masterwork
An Anatomy of Melancholy, described the "vast chaos and confusion of books" that confronted the seventeenth-century reader: "We
are oppressed with them, our eyes ache with reading, our fingers
with turning." A few years earlier, in 1600, another English writer,
Barnaby Rich, had complained, "One of the great diseases of this age
is the multitude of books that doth so overcharge the world that it
is not able to digest the abundance of idle matter that is every day
hatched and brought into the world."[46]

Ever since, we have been seeking, with mounting urgency, new
ways to bring order to the confusion of information we face every
day. For centuries, the methods of personal information management tended to be simple, manual, and idiosyncratic—filing and
shelving routines, alphabetization, annotation, notes and lists,
catalogues and concordances, rules of thumb. There were also the
more elaborate, but still largely manual, institutional mechanisms
for sorting and storing information found in libraries, universities,

and commercial and governmental bureaucracies. During the twentieth century, as the information flood swelled and data-processing
technologies advanced, the methods and tools for both personal
and institutional information management became more elaborate,
more systematic, and increasingly automated. We began to look to
the very machines that exacerbated information overload for ways to
alleviate the problem.

Vannevar Bush sounded the keynote for our modern approach
to managing information in his much-discussed article "As We May
Think," which appeared in the *Atlantic Monthly* in 1945. Bush, an
electrical engineer who had served as Franklin Roosevelt's science
adviser during World War II, worried that progress was being held
back by scientists' inability to keep abreast of information relevant
to their work. The publication of new material, he wrote, "has been
extended far beyond our present ability to make use of the record.
The summation of human experience is being expanded at a prodi-
gious rate, and the means we use for threading through the conse-
quent maze to the momentarily important item is the same as was
used in the days of square-rigged ships."

But a technological solution to the problem of information overload was, Bush argued, on the horizon: "The world has arrived at
an age of cheap complex devices of great reliability; and something
is bound to come of it." He proposed a new kind of personal cataloguing machine, called a memex, that would be useful not only to
scientists but to anyone employing "logical processes of thought."
Incorporated into a desk, the memex, Bush wrote, "is a device in
which an individual stores [in compressed form] all his books,
records, and communications, and which is mechanized so that it
may be consulted with exceeding speed and flexibility." On top of
the desk are "translucent screens" onto which are projected images
of the stored materials as well as "a keyboard" and "sets of buttons
and levers" to navigate the database. The "essential feature" of the
machine is its use of "associative indexing" to link different pieces
of information: "Any item may be caused at will to select immedi-

ately and automatically another." This process "of tying two things together is," Bush emphasized, "the important thing."[47]

With his memex, Bush anticipated both the personal computer and the hypermedia system of the World Wide Web. His article inspired many of the original developers of PC hardware and software, including such early devotees of hypertext as the famed computer engineer Douglas Engelbart and HyperCard's inventor, Bill Atkinson. But even though Bush's vision has been fulfilled to an extent beyond anything he could have imagined in his own lifetime—we are surrounded by the memex's offspring—the problem he set out to solve, information overload, has not abated. In fact, it's worse than ever. As David Levy has observed, "The development of personal digital information systems and global hypertext seems not to have solved the problem Bush identified but exacerbated it."[48]

In retrospect, the reason for the failure seems obvious. By dramatically reducing the cost of creating, storing, and sharing information, computer networks have placed far more information within our reach than we ever had access to before. And the powerful tools for discovering, filtering, and distributing information developed by companies like Google ensure that we are forever inundated by information *of immediate interest to us*—and in quantities well beyond what our brains can handle. As the technologies for data processing improve, as our tools for searching and filtering become more precise, the flood of relevant information only intensifies. More of what is of interest to us becomes visible to us. Information overload has become a permanent affliction, and our attempts to cure it just make it worse. The only way to cope is to increase our scanning and our skimming, to rely even more heavily on the wonderfully responsive machines that are the source of the problem. Today, more information is "available to us than ever before," writes Levy, "but there is less time to make use of it—and specifically to make use of it with any depth of reflection."[49] Tomorrow, the situation will be worse still.

It was once understood that the most effective filter of human thought is time. "The best rule of reading will be a method from nature, and not a mechanical one," wrote Emerson in his 1858 essay "Books." All writers must submit "their performance to the wise ear of Time, who sits and weighs, and ten years hence out of a million of pages reprints one. Again, it is judged, it is winnowed by all the winds of opinion, and what terrific selection has not passed on it, before it can be reprinted after twenty years, and reprinted after a century!"[50] We no longer have the patience to await time's slow and scrupulous winnowing. Inundated at every moment by information of immediate interest, we have little choice but to resort to automated filters, which grant their privilege, instantaneously, to the new and the popular. On the Net, the winds of opinion have become a whirlwind.

Once the train had disgorged its cargo of busy men and steamed out of the Concord station, Hawthorne tried, with little success, to return to his deep state of concentration. He glimpsed an anthill at his feet and, "like a malevolent genius," tossed a few grains of sand onto it, blocking the entrance. He watched "one of the inhabitants," returning from "some public or private business," struggle to figure out what had become of his home: "What surprise, what hurry, what confusion of mind, are expressed in his movement! How inexplicable to him must be the agency which has effected this mischief!" But Hawthorne was soon distracted from the travails of the ant. Noticing a change in the flickering pattern of shade and sun, he looked up at the clouds "scattered about the sky" and discerned in their shifting forms "the shattered ruins of a dreamer's Utopia."

IN 2007, THE American Association for the Advancement of Science invited Larry Page to deliver the keynote address at its annual conference, the country's most prestigious meeting of scientists. Page's speech was a rambling, off-the-cuff affair, but it provided a fascinating glimpse into the young entrepreneur's mind. Once again finding inspiration in an analogy, he shared with the audience his conception of human life and human intellect. "My theory is that, if you look at your programming, your DNA, it's about 600 megabytes compressed," he said, "so it's smaller than any modern operating system, smaller than Linux or Windows . . . and that includes booting up your brain, by definition. So your program algorithms probably aren't that complicated; [intelligence] is probably more about overall computation."[51]

The digital computer long ago replaced the clock, the fountain, and the factory machine as our metaphor of choice for explaining the brain's makeup and workings. We so routinely use computing terms to describe our brains that we no longer even realize we're speaking metaphorically. (I've referred to the brain's "circuits," "wiring," "inputs," and "programming" more than a few times in this book.) But Page's view is an extreme one. To him, the brain doesn't just resemble a computer; it *is* a computer. His assumption goes a long way toward explaining why Google equates intelligence with data-processing efficiency. If our brains are computers, then intelligence can be reduced to a matter of productivity—of running more bits of data more quickly through the big chip in our skull. Human intelligence becomes indistinguishable from machine intelligence.

Page has from the start viewed Google as an embryonic form of artificial intelligence. "Artificial intelligence would be the ultimate version of Google," he said in a 2000 interview, long before his company's name had become a household word. "We're nowhere near doing that now. However, we can get incrementally closer to that, and that is basically what we work on."[52] In a 2003 speech at Stanford, he went a little further in describing his company's ambition: "The ultimate search engine is something as smart as people—or smarter."[53] Sergey Brin, who says he began writing artificial-intelligence programs in middle school, shares his partner's enthusiasm for creating a true thinking machine.[54] "Certainly if you had all the world's information directly attached to your brain, or an artificial brain that was smarter than your brain, you'd be better off," he told a *Newsweek* reporter in 2004.[55] In a television interview around the same time, Brin went so far as to suggest that the "ultimate search engine" would look a lot like Stanley Kubrick's HAL. "Now, hopefully," he said, "it would never have a bug like HAL did where he killed the occupants of the spaceship. But that's what we're striving for, and I think we've made it part of the way there."[56]

The desire to build a HAL-like system of artificial intelligence may seem strange to most people. But it's a natural ambition, even an admirable one, for a pair of brilliant young computer scientists with vast quantities of cash at their disposal and a small army of programmers and engineers in their employ. A fundamentally scientific enterprise, Google is motivated by a desire to, in Eric Schmidt's words, "us[e] technology to solve problems that have never been solved before,"[57] and artificial intelligence is the hardest problem out there. Why wouldn't Brin and Page want to be the ones to crack it?

Still, their easy assumption that we'd all "be better off" if our brains were supplemented, or even replaced, by artificial intelligence is as unsettling as it is revealing. It underscores the firmness and the certainty with which Google holds to its Taylorist belief that intelligence is the output of a mechanical process, a series of discrete steps that can be isolated, measured, and optimized. "Human beings are ashamed to have been born instead of made," the twentieth-century philosopher Günther Anders once observed, and in the pronouncements of Google's founders we can sense that shame as well as the ambition it engenders.[58] In Google's world, which is the world we enter when we go online, there's little place for the pensive stillness of deep reading or the fuzzy indirection of contemplation. Ambiguity is not an opening for insight but a bug to be fixed. The human brain is just an outdated computer that needs a faster processor and a bigger hard drive—and better algorithms to steer the course of its thought.

"Everything that human beings are doing to make it easier to operate computer networks is at the same time, but for different reasons, making it easier for computer networks to operate human beings."[59]

So wrote George Dyson in *Darwin among the Machines*, his 1997 history of the pursuit of artificial intelligence. Eight years after the book came out, Dyson was invited to the Googleplex to give a talk commemorating the work of John von Neumann, the Princeton physicist who in 1945, building on the work of Alan Turing, drew up the first detailed plan for a modern computer. For Dyson, who has spent much of his life speculating about the inner lives of machines, the visit to Google must have been exhilarating. Here, after all, was a company eager to deploy its enormous resources, including many of the brightest computer scientists in the world, to create an artificial brain.

But the visit left Dyson troubled. Toward the end of an essay he wrote about the experience, he recalled a solemn warning that Turing had made in his paper "Computing Machinery and Intelligence." In our attempts to build intelligent machines, the mathematician had written, "we should not be irreverently usurping His power of creating souls, any more than we are in the procreation of children." Dyson then relayed a comment that "an unusually perceptive friend" had made after an earlier visit to the Googleplex: "I thought the coziness to be almost overwhelming. Happy Golden Retrievers running in slow motion through water sprinklers on the lawn. People waving and smiling, toys everywhere. I immediately suspected that unimaginable evil was happening somewhere in the dark corners. If the devil would come to earth, what place would be better to hide?"[60] The reaction, though obviously extreme, is understandable. With its enormous ambition, its immense bankroll, and its imperialistic designs on the world of knowledge, Google is a natural vessel for our fears as well as our hopes. "Some say Google is God," Sergey Brin has acknowledged. "Others say Google is Satan."[61]

So what *is* lurking in the dark corners of the Googleplex? Are we on the verge of the arrival of an AI? Are our silicon overlords at the door? Probably not. The first academic conference dedicated to the pursuit of artificial intelligence was held back in the summer of 1956—on the Dartmouth campus—and it seemed obvious at the time that computers would soon be able to replicate human thought. The mathematicians and engineers who convened the month-long conclave sensed that, as they wrote in a statement, "every aspect of learning or any other feature of intelligence can in principle be so precisely described that a machine can be made to simulate it."[62] It was just a matter of writing the right programs, of rendering the conscious processes of the mind into the steps of algorithms. But despite years of subsequent effort, the workings of human intelligence have eluded precise description. In the half century since the Dartmouth conference, computers have advanced at lightning speed, yet they remain, in human terms, as dumb as stumps. Our "thinking" machines still don't have the slightest idea what they're thinking. Lewis Mumford's observation that "no computer can make a new symbol out of its own resources" remains as true today as when he said it in 1967.[63]

But the AI advocates haven't given up. They've just shifted their focus. They've largely abandoned the goal of writing software programs that replicate human learning and other explicit features of intelligence. Instead, they're trying to duplicate, in the circuitry of a computer, the electrical signals that buzz among the brain's billions of neurons, in the belief that intelligence will then "emerge" from the machine as the mind emerges from the physical brain. If you can get the "overall computation" right, as Page said, then the algorithms of intelligence will write themselves. In a 1996 essay on the legacy of Kubrick's 2001, the inventor and futurist Ray Kurzweil argued that once we're able to scan a brain in sufficient detail to "ascertain the architecture of interneuronal connections in different regions," we'll be able to "design simulated neural nets that will operate in a similar fashion." Although "we can't yet build a brain like HAL's," Kurzweil concluded, "we can describe right now how we could do it."[64]

There's little reason to believe that this new approach to incubating an intelligent machine will prove any more fruitful than the old one. It, too, is built on reductive assumptions. It takes for granted

that the brain operates according to the same formal mathematical rules as a computer does—that, in other words, the brain and
the computer speak the same language. But that's a fallacy born of
our desire to explain phenomena we don't understand in terms we
do understand. John von Neumann himself warned against falling victim to this fallacy. "When we talk about mathematics," he
wrote toward the end of his life, "we may be discussing a *secondary*
language, built on the *primary* language truly used by our central
nervous system." Whatever the nervous system's language may be,
"it cannot fail to differ considerably from what we consciously and
explicitly consider as mathematics."[65]

It's also a fallacy to think that the physical brain and the thinking mind exist as separate layers in a precisely engineered "architecture." The brain and the mind, the neuroplasticity pioneers have
shown, are exquisitely intertwined, each shaping the other. As Ari
Schulman wrote in "Why Minds Are Not like Computers," a 2009
New Atlantis article, "Every indication is that, rather than a neatly
separable hierarchy like a computer, the mind is a tangled hierarchy
of organization and causation. Changes in the mind cause changes in
the brain, and vice versa." To create a computer model of the brain
that would accurately simulate the mind would require the replica-
tion of "*every* level of the brain that affects and is affected by the
mind."[66] Since we're nowhere near disentangling the brain's hierarchy, much less understanding how its levels act and interact, the
fabrication of an artificial mind is likely to remain an aspiration for
generations to come, if not forever.

Google is neither God nor Satan, and if there are shadows in the
Googleplex they're no more than the delusions of grandeur. What's
disturbing about the company's founders is not their boyish desire
to create an amazingly cool machine that will be able to outthink its
creators, but the pinched conception of the human mind that gives
rise to such a desire.

Nine

SEARCH, MEMORY

S ocrates was right. As people grew accustomed to writing down their thoughts and reading the thoughts others had written down, they became less dependent on the contents of their own memory. What once had to be stored in the head could instead be stored on tablets and scrolls or between the covers of codices. People began, as the great orator had predicted, to call things to mind not "from within themselves, but by means of external marks." The reliance on personal memory diminished further with the spread of the letterpress and the attendant expansion of publishing and literacy. Books and journals, at hand in libraries or on the shelves in private homes, became supplements to the brain's biological storehouse. People didn't have to memorize everything anymore. They could look it up.

But that wasn't the whole story. The proliferation of printed pages had another effect, which Socrates didn't foresee but may well have welcomed. Books provided people with a far greater and more diverse supply of facts, opinions, ideas, and stories than had been available before, and both the method and the culture of deep reading encouraged the commitment of printed information to memory.

In the seventh century, Isidore, the bishop of Seville, remarked how reading "the sayings" of thinkers in books "render[ed] their escape from memory less easy."[1] Because every person was free to chart his own course of reading, to define his own syllabus, individual memory became less of a socially determined construct and more the foundation of a distinctive perspective and personality. Inspired by the book, people began to see themselves as the authors of their own memories. Shakespeare has Hamlet call his memory "the book and volume of my brain."

In worrying that writing would enfeeble memory, Socrates was, as the Italian novelist and scholar Umberto Eco says, expressing "an eternal fear: the fear that a new technological achievement could abolish or destroy something that we consider precious, fruitful, something that represents for us a value in itself, and a deeply spiritual one." The fear in this case turned out to be misplaced. Books provide a supplement to memory, but they also, as Eco puts it, "challenge and improve memory; they do not narcotize it."[2]

The Dutch humanist Desiderius Erasmus, in his 1512 textbook *De Copia*, stressed the connection between memory and reading. He urged students to annotate their books, using "an appropriate little sign" to mark "occurrences of striking words, archaic or novel diction, brilliant flashes of style, adages, examples, and pithy remarks worth memorizing." He also suggested that every student and teacher keep a notebook, organized by subject, "so that whenever he lights on anything worth noting down, he may write it in the appropriate section." Transcribing the excerpts in longhand, and rehearsing them regularly, would help ensure that they remained fixed in the mind. The passages were to be viewed as "kinds of flowers," which, plucked from the pages of books, could be preserved in the pages of memory.[3]

Erasmus, who as a schoolboy had memorized great swathes of classical literature, including the complete works of the poet Horace and the playwright Terence, was not recommending memorization for memorization's sake or as a rote exercise for retaining facts. To him, memorizing was far more than a means of storage. It was the first step in a process of synthesis, a process that led to a deeper and more personal understanding of one's reading. He believed, as the classical historian Erika Rummel explains, that a person should "digest or internalize what he learns and reflect rather than slavishly reproduce the desirable qualities of the model author." Far from being a mechanical, mindless process, Erasmus's brand of memorization engaged the mind fully. It required, Rummel writes, "creativeness and judgment."[4]

Erasmus's advice echoed that of the Roman Seneca, who also used a botanical metaphor to describe the essential role that memory plays in reading and in thinking. "We should imitate bees," Seneca wrote, "and we should keep in separate compartments whatever we have collected from our diverse reading, for things conserved separately keep better. Then, diligently applying all the resources of our native talent, we should mingle all the various nectars we have tasted, and then turn them into a single sweet substance, in such a way that, even if it is apparent where it originated, it appears quite different from what it was in its original state."[5] Memory, for Seneca as for Erasmus, was as much a crucible as a container. It was more than the sum of things remembered. It was something newly made, the essence of a unique self.

Erasmus's recommendation that every reader keep a notebook of memorable quotations was widely and enthusiastically followed. Such notebooks, which came to be called "commonplace books," or just "commonplaces," became fixtures of Renaissance schooling. Every student kept one.[6] By the seventeenth century, their use had spread beyond the schoolhouse. Commonplaces were viewed as necessary tools for the cultivation of an educated mind. In 1623, Francis Bacon observed that "there can hardly be anything more useful" as "a sound help for the memory" than "a good and learned Digest of Common Places." By aiding the recording of written works in memory, he wrote, a well-maintained commonplace "supplies matter to invention."[7] Through the eighteenth century, according to American

University linguistics professor Naomi Baron, "a gentleman's commonplace book" served "both as a vehicle for and a chronicle of his intellectual development."[8]

The popularity of commonplace books ebbed as the pace of life quickened in the nineteenth century, and by the middle of the twentieth century memorization itself had begun to fall from favor. Progressive educators banished the practice from classrooms, dismissing it as a vestige of a less enlightened time. What had long been viewed as a stimulus for personal insight and creativity came to be seen as a barrier to imagination and then simply as a waste of mental energy. The introduction of new storage and recording media throughout the last century—audiotapes, videotapes, microfilm and microfiche, photocopiers, calculators, computer drives—greatly expanded the scope and availability of "artificial memory." Committing information to one's own mind seemed ever less essential. The arrival of the limitless and easily searchable data banks of the Internet brought a further shift, not just in the way we view memorization but in the way we view memory itself. The Net quickly came to be seen as a replacement for, rather than just a supplement to, personal memory. Today, people routinely talk about artificial memory as though it's indistinguishable from biological memory.

Clive Thompson, the *Wired* writer, refers to the Net as an "outboard brain" that is taking over the role previously played by inner memory. "I've almost given up making an effort to remember anything," he says, "because I can instantly retrieve the information online." He suggests that "by offloading data onto silicon, we free our own gray matter for more germanely 'human' tasks like brainstorming and daydreaming."[9] David Brooks, the popular *New York Times* columnist, makes a similar point. "I had thought that the magic of the information age was that it allowed us to know more," he writes, "but then I realized the magic of the information age is that it allows us to know less. It provides us with external cognitive servants—silicon memory systems, collaborative online filters, consumer preference algorithms and networked knowledge. We can burden these servants and liberate ourselves."[10]

Peter Suderman, who writes for the *American Scene*, argues that, with our more or less permanent connections to the Internet, "it's no longer terribly efficient to use our brains to store information." Memory, he says, should now function like a simple index, pointing us to places on the Web where we can locate the information we need at the moment we need it: "Why memorize the content of a single book when you could be using your brain to hold a quick guide to an entire library? Rather than memorize information, we now store it digitally and just remember what we stored." As the Web "teaches us to think like it does," he says, we'll end up keeping "rather little deep knowledge" in our own heads.[11] Don Tapscott, the technology writer, puts it more bluntly. Now that we can look up anything "with a click on Google," he says, "memorizing long passages or historical facts" is obsolete. Memorization is "a waste of time."[12]

Our embrace of the idea that computer databases provide an effective and even superior substitute for personal memory is not particularly surprising. It culminates a century-long shift in the popular view of the mind. As the machines we use to store data have become more voluminous, flexible, and responsive, we've grown accustomed to the blurring of artificial and biological memory. But it's an extraordinary development nonetheless. The notion that memory can be "outsourced," as Brooks puts it, would have been unthinkable at any earlier moment in our history. For the Ancient Greeks, memory was a goddess: Mnemosyne, mother of the Muses. To Augustine, it was "a vast and infinite profundity," a reflection of the power of God in man.[13] The classical view remained the common view through the Middle Ages, the Renaissance, and the Enlightenment—up to, in fact, the close of the nineteenth century. When, in an 1892 lecture before a group of teachers, William James declared that "the art of remembering is the art of thinking," he was stating the obvious.[14] Now, his words seem old-fashioned. Not only has memory lost its divinity; it's well on its way to losing its humanness. Mnemosyne has become a machine.

The shift in our view of memory is yet another manifestation of our acceptance of the metaphor that portrays the brain as a computer. If biological memory functions like a hard drive, storing bits of data in fixed locations and serving them up as inputs to the brain's calculations, then offloading that storage capacity to the Web is not just possible but, as Thompson and Brooks argue, liberating. It provides us with a much more capacious memory while clearing out space in our brains for more valuable and even "more human" computations. The analogy has a simplicity that makes it compelling, and it certainly seems more "scientific" than the suggestion that our memory is like a book of pressed flowers or the honey in a beehive's comb. But there's a problem with our new, post-Internet conception of human memory. It's wrong.

AFTER DEMONSTRATING, IN the early 1970s, that "synapses change with experience," Eric Kandel continued to probe the nervous system of the lowly sea slug for many years. The focus of his work shifted, though. He began to look beyond the neuronal triggers of simple reflex responses, such as the slug's withdrawal of its gill when touched, to the much more complicated question of how the brain stores information as memories. Kandel wanted, in particular, to shed light on one of the central and most perplexing riddles in neuroscience: how, exactly, does the brain transform fleeting short-term memories, such as the ones that enter and exit our working memory every waking moment, into the long-term memories that can last a lifetime?

Neurologists and psychologists had known since the end of the nineteenth century that our brains hold more than one kind of memory. In 1885, the German psychologist Hermann Ebbinghaus conducted an exhausting series of experiments, using himself as the sole subject, that involved memorizing two thousand nonsense words. He discovered that his ability to retain a word in memory strengthened the more times he studied the word and that it was much easier to memorize a half dozen words at a sitting than to memorize a dozen. He also found that the process of forgetting had two stages. Most of the words he studied disappeared from his memory very quickly, within an hour after he rehearsed them, but a smaller set stayed put much longer—they slipped away only gradually. The results of Ebbinghaus's tests led William James to conclude, in 1890, that memories were of two kinds: "primary memories," which evaporated from the mind soon after the event that inspired them, and "secondary memories," which the brain could hold onto indefinitely.[15]

At around the same time, studies of boxers revealed that a concussive blow to the head could bring on retrograde amnesia, erasing all memories stored during the preceding few minutes or hours while leaving older memories intact. The same phenomenon was noted in epileptics after they suffered seizures. Such observations implied that a memory, even a strong one, remains unstable for a brief period after it's formed. A certain amount of time seemed to be required for a primary, or short-term, memory to be transformed into a secondary, or long-term, one.

That hypothesis was backed up by research conducted by two other German psychologists, Georg Müller and Alfons Pilzecker, in the late 1890s. In a variation on Ebbinghaus's experiments, they asked a group of people to memorize a list of nonsense words. A day later, they tested the group and found that the subjects had no problem recalling the list. The researchers then conducted the same experiment on another group of people, but this time they had the subjects study a second list of words immediately after learning the first list. In the next day's test, this group was unable to remember the initial set of words. Müller and Pilzecker then conducted one last trial, with another twist. The third group of subjects memorized the first list of words and then, after a delay of two hours, were given the second list to study. This group, like the first, had little trouble remembering the initial list of words the next day. Müller and Pil-zecker concluded that it takes an hour or so for memories to become fixed, or "consolidated," in the brain. Short-term memories don't become long-term memories immediately, and the process of their consolidation is delicate. Any disruption, whether a jab to the head or a simple distraction, can sweep the nascent memories from the mind.[16]

Subsequent studies confirmed the existence of short-term and long-term forms of memory and provided further evidence of the importance of the consolidation phase during which the former are turned into the latter. In the 1960s, University of Pennsylvania neurologist Louis Flexner made a particularly intriguing discovery. After injecting mice with an antibiotic drug that prevented their cells from producing proteins, he found that the animals were unable to form long-term memories (about how to avoid receiving a shock while in a maze) but could continue to store short-term ones. The implication was clear: long-term memories are not just stronger forms of short-term memories. The two types of memory entail different biological processes. Storing long-term memories requires the synthesis of new proteins. Storing short-term memories does not.[17]

Inspired by the groundbreaking results of his earlier *Aplysia* experiments, Kandel recruited a team of talented researchers, including physiological psychologists and cell biologists, to help him plumb the physical workings of both short-term and long-term memory. They began to meticulously trace the course of a sea slug's neuronal signals, "one cell at a time," as the animal learned to adapt to outside stimuli such as pokes and shocks to its body.[18] They quickly confirmed what Ebbinghaus had observed: the more times an experience is repeated, the longer the memory of the experience lasts. Repetition encourages consolidation. When they examined the physiological effects of repetition on individual neurons and synapses, they discovered something amazing. Not only did the concentration of neurotransmitters in synapses change, altering the strength of the existing connections between neurons, but the neurons grew entirely new synaptic terminals. The formation of long-term memories, in other words, involves not only biochemical changes but anatomical ones. That explained, Kandel realized, why memory consolidation requires new proteins. Proteins play an essential role in producing structural changes in cells.

The anatomical alterations in the slug's relatively simple memory circuits were extensive. In one case, the researchers found that, before a long-term memory was consolidated, a particular sensory neuron had some thirteen hundred synaptic connections to about twenty-five other neurons. Only about forty percent of those connections were active—in other words, sending signals through the production of neurotransmitters. After the long-term memory had been formed, the number of synaptic connections had more than doubled, to about twenty-seven hundred, and the proportion that were active had increased from forty percent to sixty percent. The new synapses remained in place as long as the memory persisted. When the memory was allowed to fade—by discontinuing the repetition of the experience—the number of synapses eventually dropped to about fifteen hundred. The fact that, even after a memory is forgotten, the number of synapses remains a bit higher than it had been originally helps explain why it's easier to learn something a second time.

Through the new round of *Aplysia* experiments, Kandel wrote in his 2006 memoir *In Search of Memory*, "we could see for the first time that the number of synapses in the brain is not fixed—it changes with learning! Moreover, long-term memory persists for as long as the anatomical changes are maintained." The research also revealed the basic physiological difference between the two types of memory: "Short-term memory produces a change in the function of the synapse, strengthening or weakening preexisting connections; long-term memory requires anatomical changes."[19] Kandel's findings fit seamlessly with the discoveries being made about neuroplasticity by Michael Merzenich and others. Further experiments soon made it clear that the biochemical and structural changes involved in memory consolidation are not limited to slugs. They also take place in the brains of other animals, including primates.

Kandel and his colleagues had unlocked some of the secrets of memory at the cellular level. Now, they wanted to go deeper—to the molecular processes within the cells. The researchers were, as Kandel later put it, "entering completely uncharted territory."[20] They looked first at the molecular changes that occur in synapses as short-term memories are formed. They found that the process involves much more than just the transmission of a neurotransmitter—glutamate, in this case—from one neuron to another. Other types of cells, called interneurons, are also involved. The interneurons produce the neurotransmitter serotonin, which fine-tunes the synaptic connection, modulating the amount of glutamate released into the synapse. Working with the biochemists James Schwartz and Paul Greengard, Kandel discovered that the fine-tuning occurs through a series of molecular signals. The serotonin released by the interneuron binds to a receptor on the membrane of the presynaptic neuron—the neuron carrying the electric pulse—which starts a chemical reaction that leads the neuron to produce a molecule called cyclic AMP. The cyclic AMP in turn activates a protein called kinase A, a catalytic enzyme that spurs the cell to release more glutamate into the synapse, thereby strengthening the synaptic connection, prolonging the electrical activity in the linked neurons, and enabling the brain to maintain the short-term memory for seconds or minutes.

The next challenge facing Kandel was to figure out how such briefly held short-term memories could be transformed into much more permanent long-term memories. What was the molecular basis of the consolidation process? Answering that question would require him to enter the realm of genetics.

In 1983, the prestigious and well-financed Howard Hughes Medical Institute asked Kandel, together with Schwartz and the Columbia University neuroscientist Richard Axel, to head a research group in molecular cognition, based at Columbia. The group soon succeeded in harvesting neurons from larval *Aplysia* and using them to grow, as a tissue culture in the laboratory, a basic neural circuit incorporating a presynaptic neuron, a postsynaptic neuron, and the synapse between them. To mimic the action of the modulating interneurons, the scientists injected serotonin into the culture. A single squirt of serotonin, replicating a single learning experience, triggered, as expected, a release of glutamate—producing the brief strengthening of the synapse that is characteristic of short-term memory. Five separate squirts of serotonin, in contrast, strengthened the existing synapse for days and also spurred the formation of new synaptic terminals—changes characteristic of long-term memory.

What happens after repeated injections of serotonin is that the enzyme kinase A, along with another enzyme, called MAP, moves from the neuron's outer cytoplasm into its nucleus. There, kinase A activates a protein called CREB-1, which in turn switches on a set of genes that synthesize the proteins the neuron needs to grow new synaptic terminals. At the same time, MAP activates another protein, CREB-2, which switches off a set of genes that inhibit the growth of new terminals. Through a complex chemical process of cellular "marking," the resulting synaptic changes are concentrated at particular regions on the surface of the neuron and perpetuated over long periods of time. It is through this elaborate process, involving extensive chemical and genetic signals and changes, that synapses become able to hold memories over the course of days or even years. "The growth and maintenance of new synaptic terminals," writes Kandel, "makes memory persist."[21] The process also says something important about how, thanks to the plasticity of our brains, our experiences continually shape our behavior and identity: "The fact that a gene must be switched on to form long-term memory shows clearly that genes are not simply determinants of behavior but are also responsive to environmental stimulation, such as learning."[22]

THE MENTAL LIFE of a sea slug, it seems safe to say, is not particularly exciting. The memory circuits that Kandel and his team studied were simple ones. They involved the storage of what psychologists call "implicit" memories—the unconscious memories of past experiences that are recalled automatically in carrying out a reflexive action or rehearsing a learned skill. A slug calls on implicit memories when retracting its gill. A person draws on them when dribbling a basketball or riding a bike. As Kandel explains, an implicit memory "is recalled directly through performance, without any conscious effort or even awareness that we are drawing on memory."[23]

When we talk about our memories, what we're usually referring to are the "explicit" ones—the recollections of people, events, facts, ideas, feelings, and impressions that we're able to summon into the working memory of our conscious mind. Explicit memory encompasses everything that we say we "remember" about the past. Kandel refers to explicit memory as "complex memory"—and for good reason. The long-term storage of explicit memories involves all the biochemical and molecular processes of "synaptic consolidation" that play out in storing implicit memories. But it also requires a second form of consolidation, called "system consolidation," which involves concerted interactions among far-flung areas of the brain. Scientists have only recently begun to document the workings of system consolidation, and many of their findings remain tentative. What's clear, though, is that the consolidation of explicit memories involves a long and involved "conversation" between the cerebral cortex and the hippocampus.

A small, ancient part of the brain, the hippocampus lies beneath the cortex, folded deep within the medial temporal lobes. As well as being the seat of our navigational sense—it's where London cabbies store their mental maps of the city's roads—the hippocampus plays an important role in the formation and management of explicit memories. Much of the credit for the discovery of the hippocampus's connection with memory storage lies with an unfortunate man named Henry Molaison. Born in 1926, Molaison was stricken with epilepsy after suffering a severe head injury in his youth. During his adult years, he experienced increasingly debilitating grand mal seizures. The source of his affliction was eventually traced to the area of his hippocampus, and in 1953 doctors removed most of the hippocampus as well as other parts of the medial temporal lobes. The surgery cured Molaison's epilepsy, but it had an extraordinarily strange effect on his memory. His implicit memories remained intact, as did his older explicit memories. He could remember the events of his childhood in great detail. But many of his more recent explicit memories—some dating back years before the surgery—had vanished. And he was no longer able to store new explicit memories. Events slipped from his mind moments after they happened.

Molaison's experience, meticulously documented by the English psychologist Brenda Milner, suggested that the hippocampus is essential to the consolidation of new explicit memories but that after a time many of those memories come to exist independently of the hippocampus.[24] Extensive experiments over the last five decades have helped untangle this conundrum. The memory of an experience seems to be stored initially not only in the cortical regions that record the experience—the auditory cortex for a memory of a sound, the visual cortex for a memory of a sight, and so forth—but also in the hippocampus. The hippocampus provides an ideal holding place for new memories because its synapses are able to change very quickly. Over the course of a few days, through a still mysterious signaling process, the hippocampus helps stabilize the memory in the cortex, beginning its transformation from a short-term memory into a long-term one. Eventually, once the memory is fully consolidated, it appears to be erased from the hippocampus. The cortex becomes its sole holding place. Fully transferring an explicit memory from the hippocampus to the cortex is a gradual process that can take many years.[25] That's why so many of Molaison's memories disappeared along with his hippocampus.

The hippocampus seems to act as something like an orchestra conductor in directing the symphony of our conscious memory.

Beyond its involvement in fixing particular memories in the cortex, it is thought to play an important role in weaving together the various contemporaneous memories—visual, spatial, auditory, tactile, emotional—that are stored separately in the brain but that coalesce to form a single, seamless recollection of an event. Neuroscientists also theorize that the hippocampus helps link new memories with older ones, forming the rich mesh of neuronal connections that give memory its flexibility and depth. Many of the connections between memories are likely forged when we're asleep and the hippocampus is relieved of some of its other cognitive chores. As the psychiatrist Daniel Siegel explains in his book *The Developing Mind,* "Though filled with a combination of seemingly random activations, aspects of the day's experiences, and elements from the distant past, dreams may be a fundamental way in which the mind consolidates the myriad of explicit recollections into a coherent set of representations for permanent, consolidated memory."[26] When our sleep suffers, studies show, so, too, does our memory.[27]

Much remains to be learned about the workings of explicit and even implicit memory, and much of what we now know will be revised and refined through future research. But the growing body of evidence makes clear that the memory inside our heads is the product of an extraordinarily complex natural process that is, at every instant, exquisitely tuned to the unique environment in which each of us lives and the unique pattern of experiences that each of us goes through. The old botanical metaphors for memory, with their emphasis on continual, indeterminate organic growth, are, it turns out, remarkably apt. In fact, they seem to be more fitting than our new, fashionably high-tech metaphors, which equate biological memory with the precisely defined bits of digital data stored in databases and processed by computer chips. Governed by highly variable biological signals, chemical, electrical, and genetic, every aspect of human memory—the way it's formed, maintained, connected, recalled—has almost infinite gradations. Computer memory exists as simple binary bits—ones and zeros—that are processed through fixed circuits, which can be either open or closed but nothing in between.

Kobi Rosenblum, who heads the Department of Neurobiology and Ethology at the University of Haifa in Israel, has, like Eric Kandel, done extensive research on memory consolidation. One of the salient lessons to emerge from his work is how different biological memory is from computer memory. "The process of long-term memory creation in the human brain," he says, "is one of the incredible processes which is so clearly different than 'artificial brains' like those in a computer. While an artificial brain absorbs information and immediately saves it in its memory, the human brain continues to process information long after it is received, and the quality of memories depends on how the information is processed."[28] Biological memory is alive. Computer memory is not.

Those who celebrate the "outsourcing" of memory to the Web have been misled by a metaphor. They overlook the fundamentally organic nature of biological memory. What gives real memory its richness and its character, not to mention its mystery and fragility, is its contingency. It exists in time, changing as the body changes. Indeed, the very act of recalling a memory appears to restart the entire process of consolidation, including the generation of proteins to form new synaptic terminals.[29] Once we bring an explicit long-term memory back into working memory, it becomes a short-term memory again. When we reconsolidate it, it gains a new set of connections—a new context. As Joseph LeDoux explains, "The brain that does the remembering is not the brain that formed the initial memory. In order for the old memory to make sense in the current brain, the memory has to be updated."[30] Biological memory is in a perpetual state of renewal. The memory stored in a computer, by contrast, takes the form of distinct and static bits; you can move the bits from one storage drive to another as many times as you like, and they will always remain precisely as they were.

The proponents of the outsourcing idea also confuse working memory with long-term memory. When a person fails to consolidate a fact, an idea, or an experience in long-term memory, he's not "freeing up" space in his brain for other functions. In contrast to working memory, with its constrained capacity, long-term memory expands and contracts with almost unlimited elasticity, thanks to the brain's ability to grow and prune synaptic terminals and continually adjust the strength of synaptic connections. "Unlike a computer," writes Nelson Cowan, an expert on memory who teaches at the University of Missouri, "the normal human brain never reaches a point at which experiences can no longer be committed to memory; the brain cannot be full."[31] Says Torkel Klingberg, "The amount of information that can be stored in long-term memory is virtually boundless."[32] Evidence suggests, moreover, that as we build up our personal store of memories, our minds become sharper. The very act of remembering, explains clinical psychologist Sheila Crowell in *The Neurobiology of Learning*, appears to modify the brain in a way that can make it easier to learn ideas and skills in the future.[33]

We don't constrain our mental powers when we store new long-term memories. We strengthen them. With each expansion of our memory comes an enlargement of our intelligence. The Web provides a convenient and compelling supplement to personal memory, but when we start using the Web as a substitute for personal memory, bypassing the inner processes of consolidation, we risk emptying our minds of their riches.

In the 1970s, when schools began allowing students to use portable calculators, many parents objected. They worried that a reliance on the machines would weaken their children's grasp of mathematical concepts. The fears, subsequent studies showed, were largely unwarranted.[34] No longer forced to spend a lot of time on routine calculations, many students gained a deeper understanding of the principles underlying their exercises. Today, the story of the calculator is often used to support the argument that our growing dependence on online databases is benign, even liberating. In freeing us from the work of remembering, it's said, the Web allows us to devote more time to creative thought. But the parallel is flawed. The pocket calculator relieved the pressure on our working memory, letting us deploy that critical short-term store for more abstract reasoning. As the experience of math students has shown, the calculator made it easier for the brain to transfer ideas from working memory to long-term memory and encode them in the conceptual schemas that are so important to building knowledge. The Web has a very different effect. It places *more pressure* on our working memory, not only diverting resources from our higher reasoning faculties but obstructing the consolidation of long-term memories and the development of schemas. The calculator, a powerful but highly specialized tool, turned out to be an aid to memory. The Web is a technology of forgetfulness.

WHAT DETERMINES WHAT we remember and what we forget? The key to memory consolidation is attentiveness. Storing explicit memories and, equally important, forming connections between them requires strong mental concentration, amplified by repetition or by intense intellectual or emotional engagement. The sharper the attention, the sharper the memory. "For a memory to persist," writes Kandel, "the incoming information must be thoroughly and deeply processed. This is accomplished by attending to the information and associating it meaningfully and systematically with knowledge already well established in memory."[35] If we're unable to attend to the information in our working memory, the information lasts only as long as the neurons that hold it maintain their electric charge—a few seconds at best. Then it's gone, leaving little or no trace in the mind.

Attention may seem ethereal—a "ghost inside the head," as the developmental psychologist Bruce McCandliss says[36]—but it's a genuine physical state, and it produces material effects throughout the brain. Recent experiments with mice indicate that the act of pay-ing attention to an idea or an experience sets off a chain reaction that crisscrosses the brain. Conscious attention begins in the frontal lobes of the cerebral cortex, with the imposition of top-down, executive control over the mind's focus. The establishment of attention leads the neurons of the cortex to send signals to neurons in the midbrain that produce the powerful neurotransmitter dopamine. The axons of these neurons reach all the way into the hippocampus, providing a distribution channel for the neurotransmitter. Once the dopamine is funneled into the synapses of the hippocampus, it jump-starts the consolidation of explicit memory, probably by activating genes that spur the synthesis of new proteins.[37]

The influx of competing messages that we receive whenever we go online not only overloads our working memory; it makes it much harder for our frontal lobes to concentrate our attention on any one thing. The process of memory consolidation can't even get started. And, thanks once again to the plasticity of our neuronal pathways, the more we use the Web, the more we train our brain to be distracted—to process information very quickly and very efficiently but without sustained attention. That helps explain why many of us find it hard to concentrate even when we're away from our computers. Our brains become adept at forgetting, inept at remembering. Our growing dependence on the Web's information stores may in fact be the product of a self-perpetuating, self-amplifying loop. As our use of the Web makes it harder for us to lock information into our biological memory, we're forced to rely more and more on the Net's capacious and easily searchable artificial memory, even if it makes us shallower thinkers.

The changes in our brains happen automatically, outside the narrow compass of our consciousness, but that doesn't absolve us from responsibility for the choices we make. One thing that sets us apart from other animals is the command we have been granted over our attention. "'Learning how to think' really means learning how to exercise some control over *how* and *what* you think," said the novelist David Foster Wallace in a commencement address at Kenyon College in 2005. "It means being conscious and aware enough to *choose* what you pay attention to and to *choose* how you construct meaning from experience." To give up that control is to be left with "the constant gnawing sense of having had and lost some infinite thing."[38] A mentally troubled man—he would hang himself two and a half years after the speech—Wallace knew with special urgency the stakes involved in how we choose, or fail to choose, to focus our mind. We cede control over our attention at our own peril. Everything that neuroscientists have discovered about the cellular and molecular workings of the human brain underscores that point.

Socrates may have been mistaken about the effects of writing, but he was wise to warn us against taking memory's treasures for granted. His prophecy of a tool that would "implant forgetfulness" in the mind, providing "a recipe not for memory, but for reminder," has gained new currency with the coming of the Web. The prediction may turn out to have been merely premature, not wrong. Of all the sacrifices we make when we devote ourselves to the Internet as our universal medium, the greatest is likely to be the wealth of connections within our own minds. It's true that the Web is itself a network of connections, but the hyperlinks that associate bits of online data are nothing like the synapses in our brain. The Web's links are just addresses, simple software tags that direct a browser to load another discrete page of information. They have none of the organic richness or sensitivity of our synapses. The brain's connections, writes Ari Schulman, "don't merely provide *access* to a memory; they in many ways *constitute* memories."[39] The Web's connections are not *our* connections—and no matter how many hours we spend searching and surfing, they will never become our connections. When we outsource our memory to a machine, we also outsource a very important part of our intellect and even our identity. William James, in concluding his 1892 lecture on memory, said, "The connecting *is* the thinking." To which could be added, "The connecting *is* the self."

"I PROJECT THE history of the future," wrote Walt Whitman in one of the opening verses of *Leaves of Grass*. It has long been known that the culture a person is brought up in influences the content and character of that person's memory. People born into societies that celebrate individual achievement, like the United States, tend, for example, to be able to remember events from earlier in their lives than do people raised in societies that stress communal achievement, such as Korea.[40] Psychologists and anthropologists are now discovering that, as Whitman intuited, the influence goes both ways. Personal memory shapes and sustains the "collective memory" that underpins culture. What's stored in the individual mind—events, facts, concepts, skills—is more than the "representation of distinctive personhood" that constitutes the self, writes the anthropologist Pascal Boyer. It's also "the crux of cultural transmission."[41] Each of us carries and projects the history of the future. Culture is sustained in our synapses.

The offloading of memory to external data banks doesn't just threaten the depth and distinctiveness of the self. It threatens the depth and distinctiveness of the culture we all share. In a recent essay, the playwright Richard Foreman eloquently described what's at stake. "I come from a tradition of Western culture," he wrote, "in which the ideal (my ideal) was the complex, dense and 'cathedral-like' structure of the highly educated and articulate personality—a man or woman who carried inside themselves a personally constructed and unique version of the entire heritage of the West." But now, he continued, "I see within us all (myself included) the replacement of complex inner density with a new kind of self—evolving under the pressure of information overload and the technology of the 'instantly available.'" As we are drained of our "inner repertory of dense cultural inheritance," Foreman concluded, we risk turning into "pancake people—spread wide and thin as we connect with that vast network of information accessed by the mere touch of a button."[42]

Culture is more than the aggregate of what Google describes as "the world's information." It's more than what can be reduced to binary code and uploaded onto the Net. To remain vital, culture must be renewed in the minds of the members of every generation. Outsource memory, and culture withers.

a digression

on the writing of this book

I KNOW WHAT you're thinking. The very existence of this book would seem to contradict its thesis. If I'm finding it so hard to concentrate, to stay focused on a line of thought, how in the world did I manage to write a few hundred pages of at least semicoherent prose?

It wasn't easy. When I began writing *The Shallows*, toward the end of 2007, I struggled in vain to keep my mind fixed on the task. The Net provided, as always, a bounty of useful information and research tools, but its constant interruptions scattered my thoughts and words. I tended to write in disconnected spurts, the same way I wrote when blogging. It was clear that big changes were in order. In the summer of the following year, I moved with my wife from a highly connected suburb of Boston to the mountains of Colorado. There was no cell phone service at our new home, and the Internet arrived through a relatively poky DSL connection. I canceled my Twitter account, put my Facebook membership on hiatus, and mothballed my blog. I shut down my RSS reader and curtailed my skyping and instant messaging. Most important, I throttled back my e-mail application. It had long been set to check for new messages every minute. I reset it to check only once an hour, and when that still created too much of a distraction, I began keeping the program closed much of the day.

The dismantling of my online life was far from painless. For months, my synapses howled for their Net fix. I found myself sneaking clicks on the "check for new mail" button. Occasionally, I'd go on a daylong Web binge. But in time the cravings subsided, and I found myself able to type at my keyboard for hours on end or to read through a dense academic paper without my mind wandering. Some old, disused neural circuits were springing back to life, it seemed, and some of the newer, Web-wired ones were quieting down. I started to feel generally calmer and more in control of my thoughts—less like a lab rat pressing a lever and more like, well, a human being. My brain could breathe again.

My case, I realize, isn't typical. Being self-employed and of a fairly solitary nature, I have the option of disconnecting. Most people today don't. The Web is so essential to their work and social lives that even if they wanted to escape the network they could not. In a recent essay, the young novelist Benjamin Kunkel mulled over the Net's expanding hold on his waking hours: "The internet, as its proponents rightly remind us, makes for variety and convenience; it does not force anything on you. Only it turns out it doesn't feel like that at all. We don't *feel* as if we had freely chosen our online practices. We feel instead that they are habits we have helplessly picked up or that history has enforced, that we are not distributing our attention as we intend or even like to."[1]

The question, really, isn't whether people can still read or write the occasional book. Of course they can. When we begin using a new intellectual technology, we don't immediately switch from one mental mode to another. The brain isn't binary. An intellectual technology exerts its influence by shifting the emphasis of our thought. Although even the initial users of the technology can often sense the changes in their patterns of attention, cognition, and memory as their brains adapt to the new medium, the most profound shifts play out more slowly, over several generations, as the technology becomes ever more embedded in work, leisure, and education—in all the norms and practices that define a society and its culture. How is the way we read changing? How is the way we write changing? How is the way we think changing? Those are the questions we should be asking, both of ourselves and of our children.

As for me, I'm already backsliding. With the end of this book in sight, I've gone back to keeping my e-mail running all the time and I've jacked into my RSS feed again. I've been playing around with a few new social-networking services and have been posting some new entries to my blog. I recently broke down and bought a Blu-ray player with a built-in Wi-Fi connection. It lets me stream music from Pandora, movies from NetFlix, and videos from YouTube through my television and stereo. I have to confess: it's cool. I'm not sure I could live without it.

Ten

———

A THING LIKE ME

I t was one of the odder episodes in the history of computer science, yet also one of the more telling. Over the course of a few months in 1964 and 1965, Joseph Weizenbaum, a forty-one-year-old computer scientist at the Massachusetts Institute of Technology, wrote a software application for parsing written language, which he programmed to run on the university's new time-sharing system. A student, sitting at one of the system's terminals, would type a sentence into the computer, and Weizenbaum's program, following a set of simple rules about English grammar, would identify a salient word or phrase in the sentence and analyze the syntactical context in which it was used. The program would then, following another set of rules, transform the sentence into a new sentence that had the appearance of being a response to the original. The computer-generated sentence would appear almost instantly on the student's terminal, giving the illusion of a conversation.

In a January 1966 paper introducing his program, Weizenbaum provided an example of how it worked. If a person typed the sentence "I am very unhappy these days," the computer would need only know that the phrase "I am" typically comes before a description of the speaker's current situation or state of mind. The computer could then recast the sentence into the reply "How long have you been very unhappy these days?" The program worked, Weizenbaum explained, by first applying "a kind of template to the original sentence, one part of which matched the two words 'I am' and the remainder [of which] isolated the words 'very unhappy these days.'" It then used an algorithmic "reassembly kit," tailored to the template, that included a rule specifying that "any sentence of the form 'I am BLAH'" should be "transformed to 'How long have you been BLAH,' independently of the meaning of BLAH."[1]

Weizenbaum's application was a product of its time. During the 1950s and '60s, the enthusiasm for computers, software programming, and artificial intelligence gave rise not only to the idea that the human brain is a type of computer but to the sense that human language is the output of one of the algorithms running inside that computer. As David Golumbia explains in *The Cultural Logic of Computation*, a new breed of "computational linguists," led by Weizenbaum's MIT colleague Noam Chomsky, posited that the form of the "natural language" that people speak and write reflects "the operation of the computer inside the human mind that performs all linguistic operations."[2] In a 1958 article in the journal *Information and Control*, Chomsky had written that "one possible method for describing a grammar is in terms of a program for a universal Turing machine."[3] What made the computationalist theory so compelling was that it came wrapped in a seductive "penumbra of technological newness," writes Golumbia. It offered a "mechanic clarity," replacing language's human "messiness" with "a clean internal computer."[4] By reverse-engineering the way people talk, you could discover language's underlying code, which you could then replicate as software.

Weizenbaum named his program ELIZA, after Eliza Doolittle, the cockney flower girl who, in George Bernard Shaw's *Pygmalion*, learns to speak proper English under the tutelage of the haughty phonetics professor Henry Higgins. To make the simulated conversations a little more interesting, Weizenbaum also gave his artificial interlocutor a persona—that of a Rogerian psychotherapist. The choice, as he explained in his paper, was a pragmatic one. Using a technique developed by the psychologist Carl Rogers in the 1940s, Rogerian therapists pretended, in their conversations with patients, to have no understanding of the world. For the most part, they simply parroted their patients' statements back to them in the form of banal, open-ended questions or comments. Knowing that the naïveté was a pose, the patients were free to attribute to their therapists "all sorts of background knowledge, insights and reasoning ability." The Rogerian persona, Weizenbaum recognized, had "a crucial psychological utility" for ELIZA: it suggested that the program's vacuity masked some sort of real intelligence.

Weizenbaum provided a transcript of part of an exchange between ELIZA and a young woman who had been an early tester of the program (the machine's responses are in italic type):

Men are all alike.
In what way?
They're always bugging us about something or other.
Can you think of a specific example?
Well, my boyfriend made me come here.
Your boyfriend made you come here.
He says I'm depressed much of the time.
I am sorry to hear you are depressed.
It's true. I am unhappy.
Do you think coming here will help you to be not unhappy?
I need some help, that much seems certain.
What would it mean to you if you got some help?
Perhaps I could learn to get along with my mother.
Tell me more about your family.
My mother takes care of me.
Who else in your family takes care of you?
My father.
Your father.

You are like my father in some ways.

What resemblance do you see?

You are not very aggressive but I think you don't want me to notice that.

What makes you think I am not very aggressive?

You don't argue with me.

Why do you think I don't argue with you?

You are afraid of me.

Does it please you to believe I am afraid of you?

My father is afraid of everybody.[5]

In commenting on his creation, which he acknowledged to be a modest and even slightly silly contribution to the fledgling field of natural-language processing, Weizenbaum observed how easy it is for computer programmers to make machines "behave in wondrous ways, often sufficient to dazzle even the most experienced observer." But as soon as a program's "inner workings are explained in language sufficiently plain to induce understanding," he continued, "its magic crumbles away; it stands revealed as a mere collection of procedures, each quite comprehensible. The observer says to himself 'I could have written that.'" The program goes "from the shelf marked 'intelligent' to that reserved for curios."[6]

But Weizenbaum, like Henry Higgins, was soon to have his equilibrium disturbed. ELIZA quickly found fame on the MIT campus, becoming a mainstay of lectures and presentations about computing and time-sharing. It was among the first software programs able to demonstrate the power and speed of computers in a way that laymen could easily grasp. You didn't need a background in mathematics, much less computer science, to chat with ELIZA. Copies of the program proliferated at other schools as well. Then the press took notice, and ELIZA became, as Weizenbaum later put it, "a national plaything."[7] While he was surprised by the public's interest in his program, what shocked him was how quickly and deeply people using the software "became emotionally involved with the computer," talking to it as if it were an actual person. They "would, after conversing with it for a time, insist, in spite of my explanations, that the machine really understood them."[8] Even his secretary, who had watched him write the code for ELIZA "and surely knew it to be merely a computer program," was seduced. After a few moments using the software at a terminal in Weizenbaum's office, she asked the professor to leave the room because she was embarrassed by the intimacy of the conversation. "What I had not realized," said Weizenbaum, "is that extremely short exposures to a relatively simple computer program could induce powerful delusional thinking in quite normal people."[9]

Things were about to get stranger still. Distinguished psychiatrists and scientists began to suggest, with considerable enthusiasm, that the program could play a valuable role in actually treating the ill and the disturbed. In an article in the *Journal of Nervous and Mental Disease*, three prominent research psychiatrists wrote that ELIZA, with a bit of tweaking, could be "a therapeutic tool which can be made widely available to mental hospitals and psychiatric centers suffering a shortage of therapists." Thanks to the "time-sharing capabilities of modern and future computers, several hundred patients an hour could be handled by a computer system designed for this purpose." Writing in *Natural History*, the prominent astrophysicist Carl Sagan expressed equal excitement about ELIZA's potential. He foresaw the development of "a network of computer therapeutic terminals, something like arrays of large telephone booths, in which, for a few dollars a session, we would be able to talk with an attentive, tested, and largely non-directive psychotherapist."[10]

In his paper "Computing Machinery and Intelligence," Alan Turing had grappled with the question "Can machines think?" He proposed a simple experiment for judging whether a computer could be said to be intelligent, which he called "the imitation game" but which soon came to be known as the Turing test. It involved having a person, the "interrogator," sit at a computer terminal in an otherwise empty room and engage in a typed conversation with two other people, one an actual person and the other a computer pretending to be a person. If the interrogator was unable to distinguish the computer from the real person, then the computer, argued Turing, could be considered intelligent. The ability to conjure a plausible self out of words would signal the arrival of a true thinking machine.

To converse with ELIZA was to engage in a variation on the Turing test. But, as Weizenbaum was astonished to discover, the people who "talked" with his program had little interest in making rational, objective judgments about the identity of ELIZA. They *wanted* to believe that ELIZA was a thinking machine. They *wanted* to imbue ELIZA with human qualities—even when they were well aware that ELIZA was nothing more than a computer program following simple and rather obvious instructions. The Turing test, it turned out, was as much a test of the way human beings think as of the way machines think. In their *Journal of Nervous and Mental Disease* article, the three psychiatrists hadn't just suggested that ELIZA could serve as a substitute for a real therapist. They went on to argue, in circular fashion, that a psychotherapist was in essence a kind of computer: "A human therapist can be viewed as an information processor and decision maker with a set of decision rules which are closely linked to short-range and long-range goals."[11] In simulating a human being, however clumsily, ELIZA encouraged human beings to think of themselves as simulations of computers.

The reaction to the software unnerved Weizenbaum. It planted in his mind a question he had never before asked himself but that would preoccupy him for many years: "What is it about the computer that has brought the view of man as a machine to a new level of plausibility?"[12] In 1976, a decade after ELIZA's debut, he provided an answer in his book *Computer Power and Human Reason*. To understand the effects of a computer, he argued, you had to see the machine in the context of mankind's past intellectual technologies, the long succession of tools that, like the map and the clock, transformed nature and altered "man's perception of reality." Such technologies become part of "the very stuff out of which man builds his world." Once adopted, they can never be abandoned, at least not without plunging society into "great confusion and possibly utter chaos." An intellectual technology, he wrote, "becomes an indispensable component of any structure once it is so thoroughly integrated with the structure, so enmeshed in various vital substructures, that it can no longer be factored out without fatally impairing the whole structure."

That fact, almost "a tautology," helps explain how our dependence on digital computers grew steadily and seemingly inexorably after the machines were invented at the end of the Second World War. "The computer was not a prerequisite to the survival of modern society in the post-war period and beyond," Weizenbaum argued; "its enthusiastic, uncritical embrace by the most 'progressive' elements of American government, business, and industry made it a resource essential to society's survival *in the form* that the computer itself had been instrumental in shaping." He knew from his experience with time-sharing networks that the role of computers would expand beyond the automation of governmental and industrial processes. Computers would come to mediate the activities that define people's everyday lives—how they learn, how they think, how they socialize. What the history of intellectual technologies shows us, he warned, is that "the introduction of computers into some complex human activities may constitute an irreversible commitment." Our intellectual and social lives may, like our industrial routines, come to reflect the form that the computer imposes on them.[13]

What makes us most human, Weizenbaum had come to believe, is what is least computable about us—the connections between our mind and our body, the experiences that shape our memory and our thinking, our capacity for emotion and empathy. The great danger we face as we become more intimately involved with our computers—as we come to experience more of our lives through the disembodied symbols flickering across our screens—is that we'll begin to lose our humanness, to sacrifice the very qualities that separate us from machines. The only way to avoid that fate, Weizenbaum wrote, is to have the self-awareness and the courage to refuse to delegate to computers the most human of our mental activities and intellectual pursuits, particularly "tasks that demand wisdom."[14]

In addition to being a learned treatise on the workings of computers and software, Weizenbaum's book was a cri de coeur, a computer programmer's passionate and at times self-righteous examination of the limits of his profession. The book did not endear the author to his peers. After it came out, Weizenbaum was spurned as a heretic by leading computer scientists, particularly those pursuing artificial intelligence. John McCarthy, one of the organizers of the original Dartmouth AI conference, spoke for many technologists when, in a mocking review, he dismissed *Computer Power and Human Reason* as "an unreasonable book" and scolded Weizenbaum for unscientific "moralizing."[15] Outside the data-processing field, the book caused only a brief stir. It appeared just as the first personal computers were making the leap from hobbyists' workbenches to mass production. The public, primed for the start of a buying spree that would put computers into most every office, home, and school in the land, was in no mood to entertain an apostate's doubts.

WHEN A CARPENTER picks up a hammer, the hammer becomes, so far as his brain is concerned, part of his hand. When a soldier raises a pair of binoculars to his face, his brain sees through a new set of eyes, adapting instantaneously to a very different field of view. The experiments on pliers-wielding monkeys revealed how readily the plastic primate brain can incorporate tools into its sensory maps, making the artificial feel natural. In the human brain, that capacity has advanced far beyond what's seen in even our closest primate cousins. Our ability to meld with all manner of tools is one of the qualities that most distinguishes us as a species. In combination with our superior cognitive skills, it's what makes us so good at using new technologies. It's also what makes us so good at inventing them. Our brains can imagine the mechanics and the benefits of using a new device before that device even exists. The evolution of our extraordinary mental capacity to blur the boundary between the internal and the external, the body and the instrument, was, says University of Oregon neuroscientist Scott Frey, "no doubt a fundamental step in the development of technology."[16]

The tight bonds we form with our tools go both ways. Even as our technologies become extensions of ourselves, we become extensions of our technologies. When the carpenter takes his hammer into his hand, he can use that hand to do only what a hammer can do. The hand becomes an implement for pounding and pulling nails. When the soldier puts the binoculars to his eyes, he can see only what the lenses allow him to see. His field of view lengthens, but he becomes blind to what's nearby. Nietzsche's experience with his typewriter provides a particularly good illustration of the way technologies exert their influence on us. Not only did the philosopher come to imagine that his writing ball was "a thing like me"; he also sensed that he was becoming a thing like it, that his typewriter was shaping his thoughts. T. S. Eliot had a similar experience when he went from writing his poems and essays by hand to typing them. "Composing on the typewriter," he wrote in a 1916 letter to Conrad Aiken, "I find that I am sloughing off all my long sentences which I used to dote upon. Short, staccato, like modern French prose. The typewriter makes for lucidity, but I am not sure that it encourages subtlety."[17]

Every tool imposes limitations even as it opens possibilities. The more we use it, the more we mold ourselves to its form and function. That explains why, after working with a word processor for a time, I began to lose my facility for writing and editing in longhand. My experience, I later learned, was not uncommon. "People who write on a computer are often at a loss when they have to write by hand," Norman Doidge reports. Their ability "to translate thoughts into cursive writing" diminishes as they become used to tapping keys and watching letters appear as if by magic on a screen.[18] Today, with kids using keyboards and keypads from a very young age and schools discontinuing penmanship lessons, there is mounting evidence that the ability to write in cursive script is disappearing altogether from our culture. It's becoming a lost art. "We shape our tools," observed the Jesuit priest and media scholar John Culkin in 1967, "and thereafter they shape us."[19]

Marshall McLuhan, who was Culkin's intellectual mentor, elucidated the ways our technologies at once strengthen and sap us. In one of the most perceptive, if least remarked, passages in *Understanding Media*, McLuhan wrote that our tools end up "numbing" whatever part of our body they "amplify."[20] When we extend some part of ourselves artificially, we also distance ourselves from the amplified part and its natural functions. When the power loom was invented, weavers could manufacture far more cloth during the course of a workday than they'd been able to make by hand, but they sacrificed some of their manual dexterity, not to mention some of their "feel" for fabric. Their fingers, in McLuhan's terms, became numb. Farmers, similarly, lost some of their feel for the soil when they began using mechanical harrows and plows. Today's industrial farm worker, sitting in his air-conditioned cage atop a gargantuan tractor, rarely touches the soil at all—though in a single day he can till a field that his hoe-wielding forebear could not have turned in a month. When we're behind the wheel of our car, we can go a far greater distance than we could cover on foot, but we lose the walker's intimate connection to the land.

As McLuhan acknowledged, he was far from the first to observe technology's numbing effect. It's an ancient idea, one that was given perhaps its most eloquent and ominous expression by the Old Testament psalmist:

> Their idols are silver and gold,
> The work of men's hands.
> They have mouths, but they speak not;
> Eyes have they, but they see not;
> They have ears, but they hear not;
> Noses have they, but they smell not;
> They have hands, but they handle not;

Feet have they, but they walk not;

Neither speak they through their throat.

They that make them are like unto them;

So is every one that trusteth in them.

The price we pay to assume technology's power is alienation. The toll can be particularly high with our intellectual technologies. The tools of the mind amplify and in turn numb the most intimate, the most human, of our natural capacities—those for reason, perception, memory, emotion. The mechanical clock, for all the blessings it bestowed, removed us from the natural flow of time. When Lewis Mumford described how modern clocks helped "create the belief in an independent world of mathematically measurable sequences," he also stressed that, as a consequence, clocks "disassociated time from human events."[21] Weizenbaum, building on Mumford's point, argued that the conception of the world that emerged from time-keeping instruments "was and remains an impoverished version of the older one, for it rests on a rejection of those direct experiences that formed the basis for, and indeed constituted, the old reality."[22] In deciding when to eat, to work, to sleep, to wake up, we stopped listening to our senses and started obeying the clock. We became a lot more scientific, but we became a bit more mechanical as well.

Even a tool as seemingly simple and benign as the map had a numbing effect. Our ancestors' navigational skills were amplified enormously by the cartographer's art. For the first time, people could confidently traverse lands and seas they'd never seen before—an advance that spurred a history-making expansion of exploration, trade, and warfare. But their native ability to comprehend a landscape, to create a richly detailed mental map of their surroundings, weakened. The map's abstract, two-dimensional representation of space interposed itself between the map reader and his perception of the actual land. As we can infer from recent studies of the brain, the loss must have had a physical component. When people came to rely on maps rather than their own bearings, they would have experienced a diminishment of the area of their hippocampus devoted to spatial representation. The numbing would have occurred deep in their neurons.

We're likely going through another such adaptation today as we come to depend on computerized GPS devices to shepherd us around. Eleanor Maguire, the neuroscientist who led the study of the brains of London taxi drivers, worries that satellite navigation could have "a big effect" on cabbies' neurons. "We very much hope they don't start using it," she says, speaking on behalf of her team of researchers. "We believe [the hippocampal] area of the brain increased in grey matter volume because of the huge amount of data [the drivers] have to memorize. If they all start using GPS, that knowledge base will be less and possibly affect the brain changes we are seeing."[23] The cabbies would be freed from the hard work of learning the city's roads, but they would also lose the distinctive mental benefits of that training. Their brains would become less interesting.

In explaining how technologies numb the very faculties they amplify, to the point even of "autoamputation," McLuhan was not trying to romanticize society as it existed before the invention of maps or clocks or power looms. Alienation, he understood, is an inevitable by-product of the use of technology. Whenever we use a tool to exert greater control over the outside world, we change our relationship with that world. Control can be wielded only from a psychological distance. In some cases, alienation is precisely what gives a tool its value. We build houses and sew Gore-Tex jackets because we *want* to be alienated from the wind and the rain and the cold. We build public sewers because we *want* to maintain a healthy distance from our own filth. Nature isn't our enemy, but neither is it our friend. McLuhan's point was that an honest appraisal of any new technology, or of progress in general, requires a sensitivity to what's lost as well as what's gained. We shouldn't allow the glories of technology to blind our inner watchdog to the possibility that we've numbed an essential part of our self.

AS A UNIVERSAL medium, a supremely versatile extension of our senses, our cognition, and our memory, the networked computer serves as a particularly powerful neural amplifier. Its numbing effects are equally strong. Norman Doidge explains that "the computer extends the processing capabilities of our central nervous system" and in the process "also alters it." Electronic media "are so effective at altering the nervous system because they both work in similar ways and are basically compatible and easily linked." Thanks to its plasticity, the nervous system "can take advantage of this compatibility and merge with the electronic media, making a single, larger system."[24]

There's another, even deeper reason why our nervous systems are so quick to "merge" with our computers. Evolution has imbued our brains with a powerful social instinct, which, as Jason Mitchell, the head of Harvard's Social Cognition and Affective Neuroscience Laboratory, says, entails "a set of processes for inferring what those around us are thinking and feeling." Recent neuroimaging studies indicate that three highly active brain regions—one in the prefrontal cortex, one in the parietal cortex, and one at the intersection of the parietal and temporal cortices—are "specifically dedicated to the task of understanding the goings-on of other people's minds." Our innate ability for "mind reading," says Mitchell, has played an important role in the success of our species, allowing us to "coordinate large groups of people to achieve goals that individuals could not."[25] As we've entered the computer age, however, our talent for connecting with other minds has had an unintended consequence. The "chronic overactivity of those brain regions implicated in social thought" can, writes Mitchell, lead us to perceive minds where no minds exist, even in "inanimate objects." There's growing evidence, moreover, that our brains naturally mimic the states of the other minds we interact with, whether those minds are real or imagined. Such neural "mirroring" helps explain why we're so quick to attribute human characteristics to our computers and computer characteristics to ourselves—why we hear a human voice when ELIZA speaks.

Our willingness, even eagerness, to enter into what Doidge calls "a single, larger system" with our data-processing devices is an outgrowth not only of the characteristics of the digital computer as an informational medium but of the characteristics of our socially adapted brains. While this cybernetic blurring of mind and machine may allow us to carry out certain cognitive tasks far more efficiently, it poses a threat to our integrity as human beings. Even as the larger system into which our minds so readily meld is lending us its powers, it is also imposing on us its limitations. To put a new spin on Culkin's phrase, we program our computers and thereafter they program us.

Even at a practical level, the effects are not always as beneficial as we want to believe. As the many studies of hypertext and multimedia show, our ability to learn can be severely compromised when our brains become overloaded with diverse stimuli online. More information can mean less knowledge. But what about the effects of the many software tools we use? How do all the ingenious applications we depend on to find and evaluate information, form and communicate our thoughts, and carry out other cognitive chores influence what and how we learn? In 2003, a Dutch cognitive psychologist named Christof van Nimwegen began a fascinating study of computer-aided learning that a BBC writer would later call "one of the most interesting examinations of current computer use and the potential downsides of our increasing reliance on screen-based interaction with information systems."[26] Van Nimwegen had two groups of volunteers work through a tricky logic puzzle on a computer. The puzzle involved transferring colored balls between two boxes in accordance with a set of rules governing which balls could be moved at which time. One of the groups used software that had been designed to be as helpful as possible. It offered on-screen assistance during the course of solving the puzzle, providing visual cues, for instance, to highlight permitted moves. The other group used a bare-bones program, which provided no hints or other guidance.

In the early stages of solving the puzzle, the group using the helpful software made correct moves more quickly than the other group, as would be expected. But as the test proceeded, the proficiency of the members of the group using the bare-bones software increased more rapidly. In the end, those using the unhelpful program were able to solve the puzzle more quickly and with fewer wrong moves. They also reached fewer impasses—states in which no further moves were possible—than did the people using the helpful software. The findings indicated, as van Nimwegen reported, that those using the unhelpful software were better able to plan ahead and plot strategy, while those using the helpful software tended to rely on simple trial and error. Often, in fact, those with the helpful software were found "to aimlessly click around" as they tried to crack the puzzle.[27]

Eight months after the experiment, van Nimwegen reassembled the groups and had them again work on the colored-balls puzzle as well as a variation on it. He found that the people who had originally used the unhelpful software were able to solve the puzzles nearly twice as fast as those who had used the helpful software. In another test, he had a different set of volunteers use ordinary calendar software to schedule a complicated series of meetings involving overlapping groups of people. Once again, one group used helpful software that provided lots of on-screen cues, and another group used unhelpful software. The results were the same. The subjects using the unhelpful program "solved the problems with fewer superfluous moves [and] in a more straightforward manner," and they demonstrated greater "plan-based behavior" and "smarter solution paths."[28]

In his report on the research, van Nimwegen emphasized that he controlled for variations in the participants' fundamental cognitive skills. It was the differences in the design of the software that explained the differences in performance and learning. The subjects using the bare-bones software consistently demonstrated "more focus, more direct and economical solutions, better strategies, and better imprinting of knowledge." The more that people depended on explicit guidance from software programs, the less engaged they were in the task and the less they ended up learning. The findings indicate, van Nimwegen concluded, that as we "externalize" problem solving and other cognitive chores to our computers, we reduce our brain's ability "to build stable knowledge structures"—schemas, in other words—that can later "be applied in new situations."[29] A polemicist might put it more pointedly: The brighter the software, the dimmer the user.

In discussing the implications of his study, van Nimwegen suggested that programmers might want to design their software to be less helpful in order to force users to think harder. That may well be good advice, but it's hard to imagine the developers of commercial computer programs and Web applications taking it to heart. As van Nimwegen himself noted, one of the long-standing trends in software programming has been the pursuit of ever more "user-friendly" interfaces. That's particularly true on the Net. Internet companies are in fierce competition to make people's lives easier, to shift the burden of problem solving and other mental labor away from the user and onto the microprocessor. A small but telling example can be seen in the evolution of search engines. In its earliest incarnation, the Google engine was a very simple tool: you entered a keyword into the search box, and you hit the Search button. But Google, facing competition from other search engines, like Microsoft's Bing, has worked diligently to make its service ever more solicitous. Now, as soon as you enter the first letter of your keyword into the box, Google immediately suggests a list of popular search terms that begin with that letter. "Our algorithms use a wide range of information to predict the queries users are most likely to want to see," the company explains. "By suggesting more refined searches up front, [we] can make your searches more convenient and efficient."[30]

Automating cognitive processes in this way has become the modern programmer's stock-in-trade. And for good reason: people naturally seek out those software tools and Web sites that offer the most

help and the most guidance—and shun those that are difficult to
master. We *want* friendly, helpful software. Why wouldn't we? Yet
as we cede to software more of the toil of thinking, we are likely
diminishing our own brain power in subtle but meaningful ways.
When a ditchdigger trades his shovel for a backhoe, his arm muscles
weaken even as his efficiency increases. A similar trade-off may well
take place as we automate the work of the mind.

Another recent study, this one on academic research, provides
real-world evidence of the way the tools we use to sift information
online influence our mental habits and frame our thinking. James
Evans, a sociologist at the University of Chicago, assembled an enormous database on 34 million scholarly articles published in academic journals from 1945 through 2005. He analyzed the citations
included in the articles to see if patterns of citation, and hence of
research, have changed as journals have shifted from being printed
on paper to being published online. Considering how much easier
it is to search digital text than printed text, the common assumption has been that making journals available on the Net would significantly broaden the scope of scholarly research, leading to a much
more diverse set of citations. But that's not at all what Evans discovered. As more journals moved online, scholars actually cited fewer
articles than they had before. And as old issues of printed journals
were digitized and uploaded to the Web, scholars cited more recent
articles with increasing frequency. A broadening of available information led, as Evans described it, to a "narrowing of science and
scholarship."[31]

In explaining the counterintuitive findings in a 2008 *Science*
article, Evans noted that automated information-filtering tools, such
as search engines, tend to serve as amplifiers of popularity, quickly
establishing and then continually reinforcing a consensus about
what information is important and what isn't. The ease of following hyperlinks, moreover, leads online researchers to "bypass many
of the marginally related articles that print researchers" would routinely skim as they flipped through the pages of a journal or a book.

The quicker that scholars are able to "find prevailing opinion," wrote Evans, the more likely they are "to follow it, leading to more citations referencing fewer articles." Though much less efficient than searching the Web, old-fashioned library research probably served to widen scholars' horizons: "By drawing researchers through unrelated articles, print browsing and perusal may have facilitated broader comparisons and led researchers into the past."[32] The easy way may not always be the best way, but the easy way is the way our computers and search engines encourage us to take.

Before Frederick Taylor introduced his system of scientific management, the individual laborer, drawing on his training, knowledge, and experience, would make his own decisions about how he did his work. He would write his own script. After Taylor, the laborer began following a script written by someone else. The machine operator was not expected to understand how the script was constructed or the reasoning behind it; he was simply expected to obey it. The messiness that comes with individual autonomy was cleaned up, and the factory as a whole became more efficient, its output more predictable. Industry prospered. What was lost along with the messiness was personal initiative, creativity, and whim. Conscious craft turned into unconscious routine.

When we go online, we, too, are following scripts written by others—algorithmic instructions that few of us would be able to understand even if the hidden codes were revealed to us. When we search for information through Google or other search engines, we're following a script. When we look at a product recommended to us by Amazon or Netflix, we're following a script. When we choose from a list of categories to describe ourselves or our relationships on Facebook, we're following a script. These scripts can be ingenious and extraordinarily useful, as they were in the Taylorist factories, but they also mechanize the messy processes of intellectual exploration and even social attachment. As the computer programmer Thomas Lord has argued, software can end up turning the most intimate and personal of human activities into mindless "rituals" whose steps are "encoded in the logic of web pages."[33] Rather than acting according to our own knowledge and intuition, we go through the motions.

WHAT EXACTLY WAS going on in Hawthorne's head as he sat in the green seclusion of Sleepy Hollow and lost himself in contemplation? And how was it different from what was going through the minds of the city dwellers on that crowded, noisy train? A series of psychological studies over the past twenty years has revealed that after spending time in a quiet rural setting, close to nature, people exhibit greater attentiveness, stronger memory, and generally improved cognition. Their brains become both calmer and sharper. The reason, according to attention restoration theory, or ART, is that when people aren't being bombarded by external stimuli, their brains can, in effect, relax. They no longer have to tax their working memories by processing a stream of bottom-up distractions. The resulting state of contemplativeness strengthens their ability to control their mind.

The results of the most recent such study were published in *Psychological Science* at the end of 2008. A team of University of Michigan researchers, led by psychologist Marc Berman, recruited some three dozen people and subjected them to a rigorous, and mentally fatiguing, series of tests designed to measure the capacity of their working memory and their ability to exert top-down control over their attention. The subjects were then divided into two groups. Half of them spent about an hour walking through a secluded woodland park, and the other half spent an equal amount of time walking along busy downtown streets. Both groups then took the tests a second time. Spending time in the park, the researchers found, "significantly improved" people's performance on the cognitive tests, indicating a substantial increase in attentiveness. Walking in the city, by contrast, led to no improvement in test results.

The researchers then conducted a similar experiment with another set of people. Rather than taking walks between the rounds of testing, these subjects simply looked at photographs of either calm rural scenes or busy urban ones. The results were the same. The people who looked at pictures of nature scenes were able to exert substantially stronger control over their attention, while those who looked at city scenes showed no improvement in their attentiveness. "In sum," concluded the researchers, "simple and brief interactions with nature can produce marked increases in cognitive control." Spending time in the natural world seems to be of "vital importance" to "effective cognitive functioning."[34]

There is no Sleepy Hollow on the Internet, no peaceful spot where contemplativeness can work its restorative magic. There is only the endless, mesmerizing buzz of the urban street. The stimulations of the Net, like those of the city, can be invigorating and inspiring. We wouldn't want to give them up. But they are, as well, exhausting and distracting. They can easily, as Hawthorne understood, overwhelm all quieter modes of thought. One of the greatest dangers we face as we automate the work of our minds, as we cede control over the flow of our thoughts and memories to a powerful electronic system, is the one that informs the fears of both the scientist Joseph Weizenbaum and the artist Richard Foreman: a slow erosion of our humanness and our humanity.

It's not only deep thinking that requires a calm, attentive mind. It's also empathy and compassion. Psychologists have long studied how people experience fear and react to physical threats, but it's only recently that they've begun researching the sources of our nobler instincts. What they're finding is that, as Antonio Damasio, the director of USC's Brain and Creativity Institute, explains, the higher emotions emerge from neural processes that "are inherently slow."[35] In one recent experiment, Damasio and his colleagues had subjects listen to stories describing people experiencing physical or psychological pain. The subjects were then put into a magnetic resonance imaging machine and their brains were scanned as they were asked to remember the stories. The experiment revealed that while the human brain reacts very quickly to demonstrations of physical pain—when you see someone injured, the primitive pain centers in

your own brain activate almost instantaneously—the more sophisticated mental process of empathizing with psychological suffering
unfolds much more slowly. It takes time, the researchers discovered,
for the brain "to transcend immediate involvement of the body" and
begin to understand and to feel "the psychological and moral dimen-
sions of a situation."[36]

The experiment, say the scholars, indicates that the more distracted we become, the less able we are to experience the subtlest,
most distinctively human forms of empathy, compassion, and other
emotions. "For some kinds of thoughts, especially moral decision-
making about other people's social and psychological situations,
we need to allow for adequate time and reflection," cautions Mary
Helen Immordino-Yang, a member of the research team. "If things
are happening too fast, you may not ever fully experience emotions
about other people's psychological states."[37] It would be rash to jump
to the conclusion that the Internet is undermining our moral sense.
It would not be rash to suggest that as the Net reroutes our vital
paths and diminishes our capacity for contemplation, it is altering
the depth of our emotions as well as our thoughts.

There are those who are heartened by the ease with which our
minds are adapting to the Web's intellectual ethic. "Technological
progress does not reverse," writes a *Wall Street Journal* columnist,
"so the trend toward multitasking and consuming many different types of information will only continue." We need not worry,
though, because our "human software" will in time "catch up to
the machine technology that made the information abundance possible." We'll "evolve" to become more agile consumers of data.[38] The
writer of a cover story in *New York* magazine says that as we become
used to "the 21st-century task" of "flitting" among bits of online
information, "the wiring of the brain will inevitably change to deal
more efficiently with more information." We may lose our capacity
"to concentrate on a complex task from beginning to end," but in
recompense we'll gain new skills, such as the ability to "conduct 34
conversations simultaneously across six different media."[39] A promi-

nent economist writes, cheerily, that "the web allows us to borrow cognitive strengths from autism and to be better infovores."[40] An *Atlantic* author suggests that our "technology-induced ADD" may be "a short-term problem," stemming from our reliance on "cognitive habits evolved and perfected in an era of limited information flow." Developing new cognitive habits is "the only viable approach to navigating the age of constant connectivity."[41]

These writers are certainly correct in arguing that we're being molded by our new information environment. Our mental adaptability, built into the deepest workings of our brains, is a keynote of intellectual history. But if there's comfort in their reassurances, it's of a very cold sort. Adaptation leaves us better suited to our circumstances, but qualitatively it's a neutral process. What matters in the end is not our becoming but what we become. In the 1950s, Martin Heidegger observed that the looming "tide of technological revolution" could "so captivate, bewitch, dazzle, and beguile man that calculative thinking may someday come to be accepted and practiced *as the only* way of thinking." Our ability to engage in "meditative thinking," which he saw as the very essence of our humanity, might become a victim of headlong progress.[42] The tumultuous advance of technology could, like the arrival of the locomotive at the Concord station, drown out the refined perceptions, thoughts, and emotions that arise only through contemplation and reflection. The "frenziedness of technology," Heidegger wrote, threatens to "entrench itself everywhere."[43]

It may be that we are now entering the final stage of that entrenchment. We are welcoming the frenziedness into our souls.

Epilogue

HUMAN ELEMENTS

A s I was finishing this book late in 2009, I stumbled on a small story tucked away in the press. Edexcel, the largest educational testing firm in England, had announced it was introducing "artificial intelligence-based, automated marking of exam essays." The computerized grading system would "read and assess" the essays that British students write as part of a widely used test of language proficiency. A spokesman for Edexcel, which is a subsidiary of the media conglomerate Pearson, explained that the system "produced the accuracy of human markers while eliminating human elements such as tiredness and subjectivity," according to a report in the *Times Education Supplement*. A testing expert told the paper that the computerized evaluation of essays would be a mainstay of education in the future: "The uncertainty is 'when' not 'if.'"[1]

How, I wondered, would the Edexcel software discern those rare students who break from the conventions of writing not because they're incompetent but because they have a special spark of brilliance? I knew the answer: it wouldn't. Computers, as Joseph Weizenbaum pointed out, follow rules; they don't make judgments. In place of subjectivity, they give us formula. The story revealed just how prescient Weizenbaum had been when, decades ago, he warned that as we grow more accustomed to and dependent on our computers we will be tempted to entrust to them "tasks that demand wisdom." And once we do that, there will be no turning back. The software will become indispensable to those tasks.

The seductions of technology are hard to resist, and in our age of instant information the benefits of speed and efficiency can seem unalloyed, their desirability beyond debate. But I continue to hold out hope that we won't go gently into the future our computer engineers and software programmers are scripting for us. Even if we don't heed Weizenbaum's words, we owe it to ourselves to consider them, to be attentive to what we stand to lose. How sad it would be, particularly when it comes to the nurturing of our children's minds, if we were to accept without question the idea that "human elements" are outmoded and dispensable.

The Edexcel story also stirred, once again, my memory of that scene at the end of 2001. It's a scene that has haunted me ever since I first saw the film as a teenager back in the 1970s, in the midst of my analogue youth. What makes it so poignant, and so weird, is the computer's emotional response to the disassembly of its mind: its despair as one circuit after another goes dark, its childlike pleading with the astronaut—"I can feel it. I can feel it. I'm afraid"—and its final reversion to what can only be called a state of innocence. HAL's outpouring of feeling contrasts with the emotionlessness that characterizes the human figures in the film, who go about their business with an almost robotic efficiency. Their thoughts and actions feel scripted, as if they're following the steps of an algorithm. In the world of 2001, people have become so machinelike that the most human character turns out to be a machine. That's the essence of Kubrick's dark prophecy: as we come to rely on computers to mediate our understanding of the world, it is our own intelligence that flattens into artificial intelligence.

THE MOST INTERESTING THING IN THE WORLD

Marketing slogans don't normally assume the power of prophecy, but the one beamed onto a screen behind Steve Jobs the morning of January 9, 2007, was an exception. The Apple CEO, ebullient despite his long struggle with pancreatic cancer, was more than an hour into his keynote address at the annual Macworld trade show in San Francisco's Moscone Convention Center. He had reached the talk's climax—the unveiling of Apple's latest gadget, a sleek handheld computer called the iPhone. It's "a revolutionary product," he told the rapt audience. It "changes everything." And then a slide appeared with a striking, twenty-foot-tall image of the new phone. Wrapped around the picture was the prescient tagline: "Your life in your pocket."

Not even Jobs grasped how thoroughly the iPhone would recast our daily routines. "What's the killer app?" he asked the crowd. "The killer app is making calls!" To Jobs, the iPhone was a sexy, tricked-out version of the commonplace cell phone. But making calls would turn out to be the least consequential of its features. What really mattered were its powerful operating system, its versatile touch-screen, and its always-available network connection, the combination of which enabled it to run a much greater variety of software than had been possible with earlier mobile devices. The iPhone did turn out to be the future of the cell phone, but it also turned out to be the future of the personal computer. Within a few years, the smartphone had replaced the desktop and the laptop as the general public's preferred data-processing machine. By delivering a never-ending stream of information into the hands of the masses, the iPhone and its kin completed what the Internet had begun: the consolidation of communications, computing, and media into a single industry—and onto a single device.

The original iPhone went on sale in June of 2007, a few months after Jobs's speech. That was also, by happenstance, when I began the research that would culminate, three years later, with the publication of this book. Although the first edition of *The Shallows* includes several references to the iPhone and other smartphones, the story it tells is set in a time when desktops and laptops still defined people's conception of computing and framed their experience of the Internet. Even social networks like Facebook, LinkedIn, and Twitter were accessed almost exclusively through Web sites ten years ago. The apps we now download by the billions had yet to take hold.

That world feels distant now. To the young, who have grown up with phones in their hands, it must seem utterly foreign—like a world without cars or indoor plumbing. Jobs may not have anticipated the full extent of the iPhone's impact, but he got the "revolutionary" part right. Along with the attendant growth of social media, the proliferation of smartphones—more than 10 billion have been sold—has had a sweeping influence on almost every aspect of life and culture. It has given a new texture and tempo to our days. It has upset social norms and relations. It has reshaped the public square and the political arena. And it has allowed a handful of companies to hold sway over what we see, what we do, and how we express ourselves.

Jobs was mistaken, though, to suggest that the iPhone would

change everything. When it comes to how we think—the central subject of this book—smartphones and their apps have reinforced the status quo of the digital age, not upended it. They have amplified and accelerated all the psychological and cognitive trends I described in the preceding pages. A review of the sociological and scientific research that has appeared over the last decade—some of it inspired by *The Shallows*—makes that clear.

PEOPLE WERE SPENDING a lot of time looking at screens in 2010. Today, they're spending a lot more. According to the latest installment of the Nielsen Company's long-running media-use survey, the average American adult can now be found gazing into an electronic screen—television, computer, or phone—a whopping nine hours and forty-five minutes a day. That's up more than an hour and a half from five years ago.[1] Astounding as they are, the Nielsen figures appear to understate actual screen time considerably, as the company excludes from its survey "non-media" computer activities—pretty much anything that doesn't involve a Web browser or a social-media app. Once those tasks are taken into account, it becomes evident that Americans now spend at least half their waking hours looking at screens.

As screens command more of our attention, less remains for everything else. Quieter, more solitary pastimes—reading for pleasure, notably— continue to be the most vulnerable to being crowded out by digital diversions. The time Americans devote to leisure reading dropped to sixteen minutes a day in 2018 from an already paltry twenty minutes in 2008, according to the Bureau of Labor Statistics' annual time-use survey.[2] Remove the elderly from the picture, and daily reading time drops to about six minutes—less than three-quarters of an hour a week. There are still plenty of readers around, but curling up with a book is losing its place in the general culture.[3] It's becoming a quaint pursuit, like ballroom dancing or darts.

The recent rise in screen time is the direct result of the explosion in smartphone use. People who own smartphones—around eighty percent of adults and more than ninety-five percent of young adults—use their phones between four and six hours a day on average, according to the latest statistics. In one study, conducted in 2015 at the University of Lincoln in England, psychologists installed tracking software on the smartphones of twenty-three students and staff members and monitored all activity over two weeks.[4] The study participants were on their phones an average of 5.05 hours a day. That's a large amount of time, but how those hours were divided up is even more revealing. The participants used their phone eighty-five times a day on average—a number in line with data that Apple has released on iPhone use [5]—and most of those interactions were brief, the majority lasting less than thirty seconds. Phone owners tend to check their devices impulsively throughout the day, the activity logs showed, from the moment they wake up to the moment they go to bed. Smartphone use has become so "habitual," the researchers wrote, that "people have little awareness of the frequency with which they check their phone."

None of this comes as a surprise. It confirms what most of us know from experience. But the study and others like it are revelatory nonetheless. They throw into high relief the fact that the smartphone is something new in the world. Never before has a media gadget, or any other piece of technology, been so entangled with our day-to-day and even minute-by-minute existence. Contrast the smartphone with the TV. People always spent a lot of time watching television (and still do), but traditional TV viewing was concentrated at particular times—evenings, especially. It didn't extend throughout the day. People weren't carrying TVs in their pockets and pulling them out every few minutes. With smartphones, all time is prime time. Because the gadgets are always at hand—whether we're at home, at work, at school, or walking down the street—they are always intruding on our thoughts.

As smartphone use intensified after 2010, many scientists began to study the cognitive and emotional effects. Their initial findings reinforced what had already been discovered about the Internet's power to distract the mind, scatter attention, and breed anxiety. If you use your phone while doing something else—driving a car, say, or studying for an exam—your performance will suffer. But the research also revealed that our phones routinely disrupt our thinking even when we're not using them, when they're tucked away in a pocket or a purse. That's a consequence of the dozens of alerts and notifications a typical smartphone emits over the course of a day. In a 2015 Florida State University study, psychologists found that when people's phones beep or buzz while they're in the middle of a challenging task, their focus wavers and their work gets sloppier—whether they check the phone or not.[6] Another 2015 study, published in the *Journal of Computer-Mediated Communication*, showed that when people hear their phone ring but are unable to answer it, their blood pressure spikes, their pulse quickens, and their problem-solving skills weaken.[7] A 2016 experiment by a University of Virginia psychologist and two colleagues revealed that phone notifications produce symptoms of hyperactivity and absentmindedness similar to those that afflict people with attention deficit disorders.[8]

The early findings were troubling, but they only hinted at the fraught symbiosis that was developing between minds and phones. In 2017, we got a fuller picture. A team of four cognitive and behavioral psychologists, led by Adrian Ward of the University of Texas at Austin and including Kristen Duke and Ayelet Gneezy of the University of California at San Diego and Harvard's Maarten Bos, published an article called "Brain Drain" in the April issue of the *Journal of the Association for Consumer Research*.[9] It described the results of an ingenious experiment involving more than 500 undergraduates at UCSD. The students were given two standard tests of intellectual acuity. One gauged "working memory capacity," the mind's ability to focus its cognitive power on a task. The second assessed "fluid intelligence," the mind's ability to interpret and solve an unfamiliar problem. The only variable in the experiment was the location of the subjects' smartphones. Some of the students were asked to place their phones on their desks, screen side down; others were told to stow their phones in their pockets or handbags; still others were required to leave their phones in a different room. In all cases, the phones were put into do-not-disturb mode, so they would neither ring nor vibrate during the exercise.

The results were striking. In both tests, the subjects whose phones were in view posted the worst scores, while those who left their phones in a different room did the best. The students who kept their phones in their pockets or bags came out in the middle. As the phone's proximity increased, brainpower decreased. It was as if the smartphones had force fields that sapped their owners' intelligence. In subsequent interviews, nearly all the students said that their phones hadn't been a distraction—that they hadn't even thought about the devices during the experiment. They remained oblivious even as the phones muddled their thinking. A follow-up experiment, with nearly 300 participants, produced similar results. It also revealed that the more heavily the students relied on their phones in their everyday lives, the greater the cognitive penalty they suffered when their phones were nearby.

In summing up the findings, Ward and his coauthors wrote that the "integration of smartphones into daily life" appears to cause a "brain drain" that diminishes such vital mental skills as "learning, logical reasoning, abstract thought, problem solving, and creativity." Smartphones have become so tied up in our lives that, even when we're not peering or pawing at them, they tug at our attention, diverting precious cognitive resources. Just suppressing the desire to check a phone, which we do routinely and subconsciously throughout the day, can debilitate our thinking, the authors noted. The fact that most of us now habitually keep our phones "nearby and in sight" only magnifies the toll.

The "Brain Drain" study's findings are consistent with other published research. In a similar but smaller 2014 study, psychologists at the University of Southern Maine found that people who had their phones in view, albeit turned off, during two demanding tests of attention and cognition made significantly more errors than did a control group whose phones remained out of sight.[10] When the researchers gave the participants a set of easier tests, however, they found that the two groups performed about the same. That makes sense. If our minds aren't being taxed, we can spare the cognitive capacity that our phones siphon off. It's when we need to be smart that our phones dumb us down.

Because learning requires strong mental focus and exertion, students are especially susceptible to the brain-depleting effects of smartphones. A 2017 experiment at the University of Arkansas at Monticello examined how phones affected undergraduates' understanding and retention of information in a large lecture class.[11] The researchers found that students who didn't bring their phones to the classroom scored a full letter grade higher on a test of the material presented than those who had their phones with them. It didn't matter whether the students who had their phones used them or not: All of them scored equally poorly. A 2016 survey of nearly a hundred high schools in Britain found that when schools ban smartphones, students' examination scores go up substantially, with the weakest students benefiting the most.[12]

It isn't just our reasoning that takes a hit when phones are around. Our social skills and relationships appear to suffer as well. Because smartphones serve as constant reminders of all the friends we could be exchanging messages with electronically, they pull at our minds when we're talking with someone in person. Conversations become shallower and less satisfying. In a 2013 study conducted at Britain's University of Essex, 142 people were divided into pairs and asked to converse in private for ten minutes. Half talked with a phone in the room, half without. The participants were then given tests of affinity, trust, and empathy. "The mere presence of mobile phones," the researchers reported, "inhibited the development of interpersonal closeness and trust" and diminished "the extent to which individuals felt empathy and understanding from their partners." The effects were strongest when "a personally meaningful topic" was being discussed.[13] The findings were validated in a subsequent and more realistic experiment, conducted by Virginia Tech professors, that involved observing 200 people chatting in coffee shops and cafes.[14]

THE EVIDENCE THAT our phones get inside our heads with such disruptive force is unsettling. It suggests that our thoughts and feelings, far from being sheltered in our skulls, can be skewed by outside forces we're not even aware of. But however uncanny, the findings fit with what neuroscientists have discovered about the way the mind accomplishes one of its central functions: deciding what to pay attention to.

At every instant of the day, our nervous system is bombarded by stimuli that may be worthy of our attention—objects in our field of view, sounds and scents, people we know and people we don't know, ideas and memories, emotions, bodily sensations. From the near-infinite welter of possibilities, the mind has to choose a target. This enormously complicated, enormously important task—nothing so determines our thoughts and behavior as the distribution of our attention—is accomplished through a neural system called the salience network. Spanning many areas of the brain, from the subcortical limbic system that regulates basic drives and feelings to the frontal cortex that guides conscious decision-making, the salience network is, in the words of Stanford behavioral scientist Vinod Menon, "the interface of the cognitive, homeostatic, motivational, and affective systems of the human brain."[15] It is, to put it another way, the orchestrator of the self.

In selecting targets of attention, the network gives priority to four types of stimuli: those that are novel or unexpected, those that are pleasurable or otherwise rewarding, those that are personally relevant, and those that are emotionally engaging.[16] These are exactly the kinds of stimuli our smartphones supply—all the time and in abundance. Refreshing their contents continuously, our phones are fonts of new and surprising information. Our phones give us stimulation and gratification whenever we check them, triggering releases of the pleasure-producing neurotransmitter dopamine.[17] Because they are deeply personal repositories of photos and messages, our phones are always of immediate relevance to us. And our phones are emotionally charged. They send and receive signals of our social status, and they flood us with information on the people, events, and subjects we care most about. Imagine combining a mailbox, a newspaper, a TV, a radio, a photo album, a public library, a personal diary, and a boisterous party attended by everyone you know, and then compressing them all into a single, small, radiant object. That's what a smartphone represents to us.

Media and communication devices, from telephones to television sets, have always been captivating. Whether turned on or switched off, in use or idle, they promise an unending supply of interesting information and diverting experiences. By design, they seize and hold our attention in ways natural objects never could. But even in the long history of mesmerizing media, the smartphone stands out. It's an attention magnet unlike any our minds have had to grapple with before. It acts as what Ward calls a "supernormal stimulus" that is able to "hijack" attention whenever it's part of the surroundings—and it's always part of the surroundings.[18] With the smartphone, the human race has succeeded in creating the most interesting thing in the world. No wonder we can't take our minds off it.

Facebook and other social media companies have been adept at extending and exploiting the smartphone's colonization of the salience network. Building on Google's practice of exhaustive, clandestine behavioral testing, they have designed their apps to be as addictive as possible. The seemingly innocuous features we now take for granted on social media—the "like" and "heart" buttons that signal appreciation and affection, the swipe gestures that refresh the screen with new information, the "streak" counts that tally exchanges with friends, the infinite scrolls of stuff—are variations on psychological-conditioning techniques pioneered by slot-machine makers.[19] They promise emotional and social rewards, and they deliver those rewards in an unpredictable fashion. We're never sure exactly what will happen when we touch the screen, but we know we might like it. So, like compulsive gamblers, we keep coming back for more.

Social networks, Facebook's first president, Sean Parker, now admits, were designed from the start to exploit "a vulnerability in human psychology." He, his colleague Mark Zuckerberg, and other architects of the systems "understood this consciously, and we did it anyway."[20] "You don't realize it," another former Facebook executive, Chamath Palihapitiya, says, "but you are being programmed."[21] The goal of the programming is to maximize "time-on-device"—a term common to both Las Vegas and Silicon Valley. The Internet industry may have begun in idealism, but it's now powered by a manipulative and very lucrative feedback loop. The more we use our phones, the more data social-media companies amass on the way our minds respond to stimuli. They use that information to make their apps even more addictive. And the money rolls in.

Given recent advances in artificial intelligence, it doesn't take the mind of a HAL to see where this is heading. Assuming their ambitions remain unchecked, social media companies will begin using machine-learning algorithms to, as Adrian Ward puts it, "optimize for salience."[22] Through the statistical analysis of people's responses to online content, computers will be able to pinpoint the triggers of attention with a precision far beyond what Silicon Valley's army of marketers, programmers, and behavioral scientists has achieved to date. Mind control will be automated.

Steve Jobs told us we'd have our lives in our pockets. He didn't warn us about the pickpockets.

IN THE PROLOGUE to his 2000 book *From Dawn to Decadence*, the historian and social critic Jacques Barzun bemoaned the debasement of the word "culture." Through years of loose and lazy usage, it had been turned into "a piece of all-purpose jargon that covers a hodge-podge of overlapping things." Lost along the way was the term's essential meaning, which Barzun defined, simply, as "the well-furnished mind." Information of all sorts, he granted, was easier to come by in our media-saturated world, but "it may be doubted whether this bonanza will by itself cultivate the fallow mind, lift it out of day-to-day interests, and scrape it free of provincialism."[23]

It's common today, even more so than ten years ago, to think of knowledge as something that surrounds us, something we swim through and consume, like sea creatures in plankton-filled waters. The ideal of knowledge as something self-created, something woven of the facts, ideas, and experiences gathered in the individual mind, continues to recede. In the first edition of *The Shallows*, I suggested that our use of the Internet as a substitute for personal memory was misguided and dangerous. At the time, the evidence was mainly circumstantial; little research had been done. That's changed. Rigorous studies of the Web's effects on memory have been completed, and while the findings aren't definitive, they strongly suggest that our ability to form and connect memories has already been compromised.

In a 2011 study, now considered a landmark in the field, a team of researchers led by Columbia psychology professor Betsy Sparrow and including the late Harvard memory expert Daniel Wegner had people read forty brief, factual statements—"the space shuttle Columbia disintegrated during re-entry over Texas in Feb. 2003" was a typical one—then type the statements into a computer. Half the participants were told that the machine would save what they typed, and the rest were told that the statements would be erased immediately.

Afterward, the researchers asked the subjects to write down as many of the statements as they could remember. Those who believed the facts had been recorded in the computer demonstrated much weaker recall than did those who assumed the facts would not be stored. Anticipating that information will be readily available in digital form, the researchers concluded, appears to reduce the mental effort people make to remember it. Digital recording encourages neurological erasing. They dubbed this phenomenon the "Google effect," and in an article in the journal *Science* they noted its broad implications: "Since search engines are continually available to us, we may often be in a state of not feeling we need to encode the information internally. When we need it, we will look it up."[24]

People were able to look up facts long before the Internet came along—there were books, there were libraries—but it required much more time and effort. Now that it's easy to shift responsibility for memory storage and retrieval to data banks and search engines, our brains have less incentive to take on the work of remembering. Human beings are "cognitive misers," a half century of research has shown.[25] If we can offload or otherwise avoid mental work, we generally will, even when it's not in our best interest. Our phones, by giving us immediate access to pretty much every fact ever recorded, allow us to indulge our mental miserliness as never before.

Memories of facts aren't the only things that go missing. It's also memories of events. In 2014, Linda Henkel, a professor at Fairfield University in Connecticut, published the results of an experiment which revealed that when people record their experiences in digital form, they end up with foggier memories of the experiences. Henkel took a group of undergraduates into the Bellarmine Museum of Art on the Fairfield campus. She gave them digital cameras, then led them, one by one, on a tour of the museum. Along the way, she had them stop and look closely at thirty works of art—paintings, sculptures, handicrafts. In some cases, she would have a student take a photograph of an object after observing it. In other cases, she'd tell the student to set the camera aside and just look. The next day, Henkel tested the students' memories of what they had seen. She discovered they had a much tougher time recalling the works they had photographed than those they had simply observed. Even when they did remember a photographed item, they had a hazier sense of its details.[26] If prints of snapshots glued to the pages of photo albums served as aides-memoire, digital pictures stored as intangible data appear to have the opposite effect, rendering the mind less absorbent. The popular expression "pics or it didn't happen" gets it backward.

Subsequent studies have confirmed Henkel's discovery. In a series of experiments involving hundreds of subjects, Princeton psychologist Diana Tamir and three colleagues examined how people's recording of their experiences, through online comments or digital photographs, influenced memory formation in three different scenarios: watching a lecture on a computer, taking a self-guided tour of a historic building alone, and taking the same tour in the company of another person. "Media use impaired memory for both computer-based and real-world experiences, in both solo and social contexts," the researchers reported in the *Journal of Experimental Psychology*. "Creating a hard copy of an experience through media leaves only a diminished copy in our own heads."[27] With social media allowing and encouraging us to upload accounts of pretty much everything we do, this effect is now widespread. A 2017 *Frontiers in Psychology* survey of peer-reviewed research on how smartphones affect memory concluded that "when we turn to these devices, we generally learn and remember less from our experiences."[28]

There's a twist to this story. It turns out that we're not very good at distinguishing the knowledge we keep in our heads from the information we find online. As Daniel Wegner and Adrian Ward explained in a 2013 *Scientific American* article, when people call up information through their phones or other computers, they often end up suffering delusions of intelligence. They feel as though "their *own* mental capacities" had generated the information, not their devices.[29] Several studies, including an extensive series of experiments at Yale, have documented this "misattribution" phenomenon, revealing that as people gather information online, they come to believe they're smarter and more knowledgeable than they actually are.[30] "The advent of the 'information age' seems to have created a generation of people who feel they know more than ever before," Wegner and Ward concluded, even though "they may know ever less about the world around them."

That unhappy insight probably helps explain society's current gullibility crisis, with its attendant plague of propaganda, dogma, and venom. If your phone has blunted your powers of discernment, you'll believe anything it tells you. And you won't hesitate to share deceptive information with others. A 2018 MIT study of message threads on Twitter, spanning more than 4.5 million tweets posted over ten years, found that fabricated or otherwise misleading stories are 70 percent more likely to be retweeted than factual ones. While accurate stories rarely reach more than a thousand people, fake reports routinely reach tens of thousands. We want to blame algorithms and bots for the circulation of lies online, but the real culprits, the researchers discovered, are people: "False news spreads farther, faster, deeper, and more broadly than the truth because humans, not robots, are more likely to spread it."[31] The technology we assumed would enlarge us has made us smaller.

Data, the novelist and critic Cynthia Ozick once wrote, is "memory without history."[32] Her observation points to the fundamental problem with allowing smartphones and the companies that program them to commandeer our brains. When we constrict our capacity for reasoning and recall, or transfer those skills to a machine or a corporation, we sacrifice the ability to turn information into knowledge. We get the data but lose the meaning. Barring a cultural course correction, that may be the Internet's most enduring legacy.

Notes

Introduction to the Second Edition

1. Janna Quitney Anderson and Lee Rainie, *The Future of the Internet* (Washington, DC: Pew Research Center, 2010), 6.

Prologue   THE WATCHDOG AND THE THIEF

1. Marshall McLuhan, *Understanding Media: The Extensions of Man*, critical ed., ed. W. Terrence Gordon (Corte Madera, CA: Gingko, 2003), 5.
2. Ibid., 30.
3. Ibid., 31.
4. Ibid., 23.
5. Ibid., 31.
6. David Thomson, *Have You Seen?: A Personal Introduction to 1,000 Films* (New York: Knopf, 2008), 149.

One   HAL AND ME

1. Heather Pringle, "Is Google Making Archaeologists Smarter?," *Beyond Stone & Bone* blog (Archaeological Institute of America), February 27, 2009, http://archaeology.org/blog/?p=332.
2. Clive Thompson, "Your Outboard Brain Knows All," *Wired*, October 2007.
3. Scott Karp, "The Evolution from Linear Thought to Networked Thought," *Publishing 2.0* blog, February 9, 2008, http://publishing2.com/2008/02/09/the-evolution-from-linear-thought-to-networked-thought.
4. Bruce Friedman, "How Google Is Changing Our Information-Seeking Behavior," *Lab Soft News* blog, February 6, 2008, http://labsoftnews.typepad.com/lab_soft_news/2008/02/how-google-is-c.html.

5. Philip Davis, "Is Google Making Us Stupid? Nope!" *The Scholarly Kitchen* blog, June 16, 2008, http://scholarlykitchen.sspnet.org/2008/06/16/is-google-making-us-stupid-nope.

6. Scott Karp, "Connecting the Dots of the Web Revolution," *Publishing 2.0* blog, June 17, 2008, http://publishing2.com/2008/06/17/connecting-the-dots-of-the-web-revolution.

7. Davis, "Is Google Making Us Stupid? Nope!"

8. Don Tapscott, "How Digital Technology Has Changed the Brain," *Business-Week Online*, November 10, 2008, www.businessweek.com/technology/content/nov2008/tc2008117_034517.htm.

9. Don Tapscott, "How to Teach and Manage 'Generation Net,'" *BusinessWeek Online*, November 30, 2008, www.businessweek.com/technology/content/nov2008/tc20081130_713563.htm.

10. Quoted in Naomi S. Baron, *Always On: Language in an Online and Mobile World* (Oxford: Oxford University Press, 2008), 204.

11. John Battelle, "Google: Making Nick Carr Stupid, but It's Made This Guy Smarter," *John Battelle's Searchblog*, June 10, 2008, http://battellemedia.com/archives/004494.php.

12. John G. Kemeny, *Man and the Computer* (New York: Scribner, 1972), 21.

13. Gary Wolfe, "The (Second Phase of the) Revolution Has Begun," *Wired*, October 1994.

Two THE VITAL PATHS

1. Sverre Avnskog, "Who Was Rasmus Malling-Hansen?," Malling-Hansen Society, 2006, www.malling-hansen.org/fileadmin/biography/biography.pdf.

2. The story of Nietzsche and his typewriter draws from Friedrich A. Kittler, *Gramophone, Film, Typewriter* (Stanford: Stanford University Press, 1999), 200–203; J. C. Nyíri, "Thinking with a Word Processor," in *Philosophy and the Cognitive Sciences*, ed. R. Casati (Vienna: Hölder-Pichler-Tempsky, 1994), 63–74; Christian J. Emden, *Nietzsche on Language, Consciousness, and the Body* (Champaign: University of Illinois Press, 2005), 27–29; and Curtis Cate, *Friedrich Nietzsche* (Woodstock, NY: Overlook, 2005), 315–18.

3. Joseph LeDoux, *Synaptic Self: How Our Brains Become Who We Are* (New York: Penguin, 2002), 38–39.

4. In addition to the 100 billion neurons in our brains, there are about a trillion glial cells, or glia. It was once assumed that glia were inert, essentially providing padding to the neurons. (*Glia* means "glue" in Greek.) Over the

last two decades, however, neuroscientists have found clues that glia may play important roles in the brain's functioning. A particularly abundant kind of glial cell, called an astrocyte, appears to release calcium atoms and produce neurotransmitters in response to signals from other cells. Further discoveries about glia may deepen our understanding of the brain's workings. For a good overview, see Carl Zimmer, "The Dark Matter of the Human Brain," *Discover*, September 2009.

5. J. Z. Young, *Doubt and Certainty in Science: A Biologist's Reflections on the Brain* (London: Oxford University Press, 1951), 36.

6. William James, *The Principles of Psychology*, vol. 1 (New York: Holt, 1890), 104–6. Translation of Dumont's essay is from James E. Black and William T. Greenough, "Induction of Pattern in Neural Structure by Experience: Implications for Cognitive Development," in *Advances in Developmental Psychology*, vol. 4, ed. Michael E. Lamb, Ann L. Brown, and Barbara Rogoff (Hillsdale, NJ: Erlbaum, 1986), 1.

7. See Norman Doidge, *The Brain That Changes Itself: Stories of Personal Triumph from the Frontiers of Brain Science* (New York: Penguin, 2007), 223.

8. Quoted in Jeffrey M. Schwartz and Sharon Begley, *The Mind and the Brain: Neuroplasticity and the Power of Mental Force* (New York: Harper Perennial, 2003), 130.

9. Quoted in Doidge, *Brain That Changes Itself*, 201.

10. The Nobel laureate David Hubel made this remark to the neurosurgeon Joseph Boden, report Schwartz and Begley in *Mind and the Brain*, 25.

11. Doidge, *Brain That Changes Itself*, xviii.

12. A video of the debate between Mailer and McLuhan can be seen at Google Videos: http://video.google.com/videoplay?docid=5470443898801103219.

13. Schwartz and Begley, *Mind and the Brain*, 175.

14. R. L. Paul, H. Goodman, and M. Merzenich, "Alterations in Mechanoreceptor Input to Brodmann's Areas 1 and 3 of the Postcentral Hand Area of *Macaca mulatta* after Nerve Section and Regeneration," *Brain Research*, 39, no. 1 (April 1972): 1–19.

15. Quoted in Schwartz and Begley, *Mind and the Brain*, 177.

16. James Olds, interview with the author, February 1, 2008.

17. Graham Lawton, "Is It Worth Going to the Mind Gym?," *New Scientist*, January 12, 2008.

18. The workings of synapses are extraordinarily complicated, influenced by a wide array of chemicals including transmitters like glutamate (which encourages the transfer of electrical signals between neurons) and GABA (gamma-aminobutyric acid, which inhibits the transfer of the signals) and

various modulators, like serotonin, dopamine, testosterone, and estrogen, that alter the efficacy of the transmitters. In rare cases, the membranes of neurons fuse, allowing electrical signals to pass without the mediation of synapses. See LeDoux, *Synaptic Self*, particularly 49–64.

19. Eric R. Kandel, *In Search of Memory: The Emergence of a New Science of Mind* (New York: Norton, 2006), 198–207. See also Bruce E. Wexler, *Brain and Culture: Neurobiology, Ideology, and Social Change* (Cambridge, MA: MIT Press, 2006), 27–29.

20. Kandel, *In Search of Memory*, 202–3.

21. LeDoux, *Synaptic Self*, 3.

22. The use of the visual cortex in reading Braille was documented in an experiment undertaken by Alvaro Pascual-Leone in 1993. See Doidge, *Brain That Changes Itself*, 200.

23. McGovern Institute for Brain Research, "What Drives Brain Changes in Macular Degeneration?," press release, March 4, 2009.

24. Sandra Blakesley, "Missing Limbs, Still Atingle, Are Clues to Changes in the Brain," *New York Times*, November 10, 1992.

25. In some of the most promising experimental treatments for Alzheimer's disease, currently being tested with considerable success in mice, drugs are used to promote plastic synaptic changes that strengthen memory formation. See J.-S. Guan, S. J. Haggarty, E. Giacometti, et al., "HDAC2 Negatively Regulates Memory Formation and Synaptic Plasticity," *Nature*, 459 (May 7, 2009): 55–60.

26. Mark Hallett, "Neuroplasticity and Rehabilitation," *Journal of Rehabilitation Research and Development*, 42, no. 4 (July–August 2005): xvii–xxii.

27. A. Pascual-Leone, A. Amedi, F. Fregni, and L. B. Merabet, "The Plastic Human Brain Cortex," *Annual Review of Neuroscience*, 28 (2005): 377–401.

28. David J. Buller, *Adapting Minds: Evolutionary Psychology and the Persistent Quest for Human Nature* (Cambridge, MA: MIT Press, 2005), 136–42.

29. M. A. Umiltà, L. Escola, I. Instkirveli, et al., "When Pliers Become Fingers in the Monkey Motor System," *Proceedings of the National Academy of Sciences*, 105, no. 6 (February 12, 2008): 2209–13. See also Angelo Maravita and Atsushi Iriki, "Tools for the Body (Schema)," *Trends in Cognitive Science*, 8, no. 2 (February 2004): 79–86.

30. E. A. Maguire, D. G. Gadian, I. S. Johnsrude, et al., "Navigation-Related Structural Change in the Hippocampi of Taxi Drivers," *Proceedings of the National Academy of Sciences*, 97, no. 8 (April 11, 2000): 4398–403. See also E. A. Maguire, H. J. Spiers, C. D. Good, et al., "Navigation Expertise and the Human Hippocampus: A Structural Brain Imaging Analysis," *Hip-*

pocampus, 13, no. 2 (2003): 250–59; and Alex Hutchinson, "Global Impositioning Systems," *Walrus*, November 2009.

31. A. Pascual-Leone, D. Nguyet, L. G. Cohen, et al., "Modulation of Muscle Responses Evoked by Transcranial Magnetic Stimulation during the Acquisition of New Fine Motor Skills," *Journal of Neurophysiology*, 74, no. 3 (1995): 1037–45. See also Doidge, *Brain That Changes Itself*, 200–202.

32. Michael Greenberg, "Just Remember This," *New York Review of Books*, December 4, 2008.

33. Doidge, *Brain That Changes Itself*, 317.

34. Ibid., 108.

35. Pascual-Leone et al., "Plastic Human Brain Cortex." See also Sharon Begley, *Train Your Mind, Change Your Brain: How a New Science Reveals Our Extraordinary Potential to Transform Ourselves* (New York: Ballantine, 2007), 244.

36. Doidge, *Brain That Changes Itself*, 59.

37. Schwartz and Begley, *Mind and the Brain*, 201.

a digression ON WHAT THE BRAIN THINKS ABOUT WHEN IT THINKS ABOUT ITSELF

1. Quotations from Aristotle's *The Parts of Animals* are from William Ogle's much-reproduced translation.

2. Robert L. Martensen, *The Brain Takes Shape: An Early History* (New York: Oxford University Press, 2004), 50.

3. René Descartes, *The World and Other Writings*, ed. Stephen Gaukroger (Cambridge: Cambridge University Press, 1998), 106–40.

4. Martensen, *Brain Takes Shape*, 66.

Three TOOLS OF THE MIND

1. Vincent Virga and the Library of Congress, *Cartographia* (New York: Little, Brown, 2007), 5.

2. Ibid.

3. Arthur H. Robinson, *Early Thematic Mapping in the History of Cartography* (Chicago: University of Chicago Press, 1982), 1.

4. Jacques Le Goff, *Time, Work, and Culture in the Middle Ages* (Chicago: University of Chicago Press, 1980), 44.

5. David S. Landes, *Revolution in Time: Clocks and the Making of the Modern World* (Cambridge, MA: Harvard University Press, 2000), 76.

6. Lynn White Jr., *Medieval Technology and Social Change* (New York: Oxford University Press, 1964), 124.

7. Landes, *Revolution in Time*, 92–93.

8. Lewis Mumford, *Technics and Civilization* (New York: Harcourt Brace, 1963), 15. The distinguished computer scientist Danny Hillis notes that "the computer, with its mechanistic playing out of predetermined rules, is the direct descendant of the clock." W. Daniel Hillis, "The Clock," in *The Greatest Inventions of the Past 2,000 Years*, ed. John Brockman (New York: Simon & Schuster, 2000), 141.

9. Karl Marx, *The Poverty of Philosophy* (New York: Cosimo, 2008), 119.

10. Ralph Waldo Emerson, "Ode, Inscribed to W. H. Channing," in *Collected Poems and Translations* (New York: Library of America, 1994), 63.

11. Marshall McLuhan, *Understanding Media: The Extensions of Man*, critical ed., ed. W. Terrence Gordon (Corte Madera, CA: Gingko, 2003), 68. For a more recent expression of this view, see Kevin Kelly, "Humans Are the Sex Organs of Technology," *The Technium* blog, February 16, 2007, www .kk.org/thetechnium/archives/2007/02/humans_are_the.php.

12. James W. Carey, *Communication as Culture: Essays on Media and Society* (New York: Routledge, 2008), 107.

13. Langdon Winner, "Technologies as Forms of Life," in *Readings in the Philosophy of Technology*, ed. David M. Kaplan (Lanham, MD: Rowman & Littlefield, 2004), 105.

14. Ralph Waldo Emerson, "Intellect," in *Emerson: Essays and Lectures* (New York: Library of America, 1983), 417.

15. See Maryanne Wolf, *Proust and the Squid: The Story and Science of the Reading Brain* (New York: Harper, 2007), 217.

16. H. G. Wells, *World Brain* (New York: Doubleday, Doran, 1938), vii.

17. René Descartes, *The Philosophical Writings of Descartes*, vol. 3, *The Correspondence* (Cambridge: Cambridge University Press, 1991), 304.

18. Walter J. Ong, *Orality and Literacy* (New York: Routledge, 2002), 82.

19. F. Ostrosky-Solís, Miguel Arellano García, and Martha Pérez, "Can Learning to Read and Write Change the Brain Organization? An Electrophysiological Study," *International Journal of Psychology*, 39, no. 1 (2004): 27–35.

20. Wolf, *Proust and the Squid*, 36.

21. E. Paulesu, J.-F. Démonet, F. Fazio, et al., "Dyslexia: Cultural Diversity and Biological Unity," *Science*, 291 (March 16, 2001): 2165–67. See also Mag-

gie Jackson, *Distracted: The Erosion of Attention and the Coming Dark Age* (Amherst, NY: Prometheus, 2008), 168–69.

22. Wolf, *Proust and the Squid*, 29.

23. Ibid., 34.

24. Ibid., 60–65.

25. Quotations from *Phaedrus* are taken from the popular translations by Reginald Hackforth and Benjamin Jowett.

26. Eric A. Havelock, *Preface to Plato* (Cambridge, MA: Harvard University Press, 1963), 41.

27. Ong, *Orality and Literacy*, 80.

28. See Ong, *Orality and Literacy*, 33.

29. Ibid., 34.

30. Eric A. Havelock, *The Muse Learns to Write: Reflections on Orality and Literacy from Antiquity to the Present* (New Haven, CT: Yale University Press, 1986), 74.

31. McLuhan, *Understanding Media*, 112–13.

32. Ibid., 120.

33. Ong, *Orality and Literacy*, 14–15.

34. Ibid., 82.

Four THE DEEPENING PAGE

1. Saint Augustine, *Confessions*, trans. R. S. Pine-Coffin (London: Penguin, 1961), 114.

2. Paul Saenger, *Space between Words: The Origins of Silent Reading* (Palo Alto, CA: Stanford University Press, 1997), 14.

3. Ibid., 7.

4. Ibid., 11.

5. Ibid., 15.

6. Maryanne Wolf, *Proust and the Squid: The Story and Science of the Reading Brain* (New York: Harper, 2007), 142–46.

7. Saenger, *Space between Words*, 13.

8. Charles E. Connor, Howard E. Egeth, and Steven Yantis, "Visual Attention: Bottom-Up versus Top-Down," *Cognitive Biology*, 14 (October 5, 2004): 850–52.

9. Maya Pines, "Sensing Change in the Environment," in *Seeing, Hearing, and Smelling in the World: A Report from the Howard Hughes Medical Institute*, February 1995, www.hhmi.org/senses/a120.html.

10. The brain's maintenance of top-down control over attention seems to require the synchronized firing of neurons in the prefrontal cortex. "It takes a lot of your prefrontal brain power to force yourself not to process a strong [distracting] input," says MIT neuroscientist Robert Desimone. See John Tierney, "Ear Plugs to Lasers: The Science of Concentration," *New York Times*, May 5, 2009.

11. Vaughan Bell, "The Myth of the Concentration Oasis," *Mind Hacks* blog, February 11, 2009, www.mindhacks.com/blog/2009/02/the_myth_of_the_conc.html.

12. Quoted in Alberto Manguel, *A History of Reading* (New York: Viking, 1996), 49. Early Christians practiced a religious form of Bible reading called *lectio divina*, or holy reading. Deeply meditative reading was seen as a way to approach the divine.

13. See Saenger, *Space between Words*, 249–50.

14. Ibid., 258. Walter J. Ong notes that editorial intensity increased further as the publishing business grew more sophisticated: "Print involves many persons besides the author in the production of a work—publishers, literary agents, publishers' readers, copy editors and others. Before as well as after scrutiny by such persons, writing for print often calls for painstaking revisions by the author of an order of magnitude virtually unknown in a manuscript culture." Ong, *Orality and Literacy* (New York: Routledge, 2002), 122.

15. Saenger, *Space between Words*, 259–60.

16. See Christopher de Hamel, "Putting a Price on It," introduction to Michael Olmert, *The Smithsonian Book of Books* (Washington, DC: Smithsonian Books, 1992), 10.

17. James Carroll, "Silent Reading in Public Life," *Boston Globe*, February 12, 2007.

18. Gutenberg was not the first to invent movable type. Around 1050, a Chinese craftsman named Pi Sheng began molding Chinese logographs out of small bits of clay. The clay type was used to print pages through hand-rubbing, the same method used to make prints from woodblocks. Because the Chinese didn't invent a printing press (perhaps because the large number of logographic symbols made the machine impractical), they were unable to mass-produce the prints, and Pi Sheng's movable type remained of limited use. See Olmert, *Smithsonian Book of Books*, 65.

19. See Frederick G. Kilgour, *The Evolution of the Book* (New York: Oxford University Press, 1998), 84–93.

20. Francis Bacon, *The New Organon*, ed. Lisa Jardine and Michael Silverthorne (Cambridge: Cambridge University Press, 2000), 100.

21. Elizabeth L. Eisenstein, *The Printing Press as an Agent of Change*, one-volume paperback ed. (Cambridge: Cambridge University Press, 1980), 46.

22. Michael Clapham, "Printing," in *A History of Technology*, vol. 3, *From the Renaissance to the Industrial Revolution, c. 1500–c. 1750*, ed. Charles Singer et al. (London: Oxford University Press, 1957), 37.

23. Eisenstein, *Printing Press as an Agent of Change*, 50.

24. Ibid., 49.

25. François Rabelais, *Gargantua and Pantagruel*, trans. Sir Thomas Urquhart and Pierre Le Motteux (New York: Barnes & Noble, 2005), 161.

26. Eisenstein, *Printing Press as an Agent of Change*, 72.

27. Quoted in Joad Raymond, *The Invention of the Newspaper: English Newsbooks, 1641–1649* (Oxford: Oxford University Press, 2005), 187.

28. See Olmert, *Smithsonian Book of Books*, 301.

29. Eisenstein, *Printing Press as an Agent of Change*, 130.

30. Notes Eisenstein, "Reading out loud to hearing publics not only persisted after printing but was, indeed, facilitated by the new abundance of texts." Elizabeth L. Eisenstein, *The Printing Revolution in Early Modern Europe*, 2nd ed. (New York: Cambridge University Press, 2005), 328.

31. J. Z. Young, *Doubt and Certainty in Science: A Biologist's Reflections on the Brain* (London: Oxford University Press, 1951), 101.

32. Books also introduced a new set of tools for organizing and conveying information. As Jack Goody has shown, lists, tables, formulas, and recipes became commonplace as books proliferated. Such literary devices further deepened our thinking, providing ways to classify and explain phenomena with ever-greater precision. Goody writes that "it does not require much reflection upon the contents of a book to realize the transformation in communication that writing has made, not simply in a mechanical sense, but in a cognitive one, what we can do with our minds and what our minds can do with us." Goody, *The Domestication of the Savage Mind* (Cambridge: Cambridge University Press, 1977), 160.

33. Darnton points out that the radically democratic and meritocratic Republic of Letters was an ideal that would never be fully realized, but as an ideal it had great force in shaping people's conception of themselves and their culture. Robert Darnton, "Google and the Future of Books," *New York Review of Books*, February 12, 2009.

34. David M. Levy, *Scrolling Forward: Making Sense of Documents in the Digital Age* (New York: Arcade, 2001), 104. The italics are Levy's.

35. Nicole K. Speer, Jeremy R. Reynolds, Khena M. Swallow, and Jeffrey M. Zacks, "Reading Stories Activates Neural Representations of Visual and

Motor Experiences," *Psychological Science*, 20, no. 8 (2009): 989–99. Gerry Everding, "Readers Build Vivid Mental Simulations of Narrative Situations, Brain Scans Suggest," Washington University (St. Louis) Web site, January 26, 2009, http://news-info.wustl.edu/tips/page/normal/13325.html.

36. Ralph Waldo Emerson, "Thoughts on Modern Literature," *Dial*, October 1840.

37. Ong, *Orality and Literacy*, 8.

38. Eisenstein, *Printing Press as an Agent of Change*, 152.

39. Wolf, *Proust and the Squid*, 217–18.

40. Some people have suggested that communication on the Internet, which tends to be brief, informal, and conversational, will return us to an oral culture. But that seems unlikely for many reasons, the most important being that the communication does not take place in person, as it does in oral cultures, but rather through a technological intermediary. Digital messages are disembodied. "The oral word," wrote Walter Ong, "never exists in a simply verbal context, as a written word does. Spoken words are always modifications of a total, existential situation, which always engages the body. Bodily activity beyond mere vocalization is not adventitious or contrived, but is natural and even inevitable." Ong, *Orality and Literacy*, 67–68.

41. Ibid., 80.

a digression
ON LEE DE FOREST AND HIS AMAZING AUDION

1. Public Broadcasting System, "A Science Odyssey: People and Discoveries: Lee de Forest," undated, www.pbs.org/wgbh/aso/databank/entries/btfore. html. For an excellent review of de Forest's early career and accomplishments, see Hugh G. J. Aitken, *The Continuous Wave: Technology and American Radio, 1900–1932* (Princeton, NJ: Princeton University Press, 1985), 162–249. For de Forest's own take on his life, see *Father of the Radio: The Autobiography of Lee de Forest* (Chicago: Wilcox & Follett, 1950).

2. Aitken, *Continuous Wave*, 217.

3. Lee de Forest, "Dawn of the Electronic Age," *Popular Mechanics*, January 1952.

Five A MEDIUM OF THE MOST GENERAL NATURE

1. Andrew Hodges, "Alan Turing," in *The Stanford Encyclopedia of Philosophy*, Fall 2008 ed., ed. Edward N. Zalta, http://plato.stanford.edu/archives/fall2008/entries/turing.

2. Alan Turing, "On Computable Numbers, with an Application to the Entsheidungsproblem," *Proceedings of the London Mathematical Society*, 42, no. 1 (1937): 230–65.

3. Alan Turing, "Computing Machinery and Intelligence," *Mind*, 59 (October 1950): 433–60.

4. George B. Dyson, *Darwin among the Machines: The Evolution of Global Intelligence* (New York: Addison-Wesley, 1997), 40.

5. Nicholas G. Carr, *Does IT Matter?* (Boston: Harvard Business School Press, 2004), 79.

6. K. G. Coffman and A. M. Odlyzko, "Growth of the Internet," AT&T Labs monograph, July 6, 2001, www.dtc.umn.edu/%7Eodlyzko/doc/oft.internet.growth.pdf.

7. Forrester Research, "Consumers' Behavior Online: A 2007 Deep Dive," April 18, 2008, www.forrester.com/Research/Document/0,7211,45266,00.html.

8. Forrester Research, "Consumer Behavior Online: A 2009 Deep Dive," July 27, 2009, www.forrester.com/Research/Document/0,7211,54327,00.html.

9. Nielsen Company, "Time Spent Online among Kids Increases 63 Percent in the Last Five Years, According to Nielsen," media alert, July 6, 2009, www.nielsen-online.com/pr/pr_090706.pdf.

10. Forrester Research, "A Deep Dive into European Consumers' Online Behavior, 2009," August 13, 2009, www.forrester.com/Research/Document/0,7211,54524,00.html.

11. TNS Global, "Digital World, Digital Life," December 2008, www.tnsglobal.com/_assets/files/TNS_Market_Research_Digital_World_Digital_Life.pdf.

12. Nielsen Company, "Texting Now More Popular than Calling," news release, September 22, 2008, www.nielsenmobile.com/html/press%20releases/TextsVersusCalls.html; Eric Zeman, "U.S. Teens Sent 2,272 Text Messages per Month in 4Q08," *Over the Air* blog (*InformationWeek*), May 26, 2009, www.informationweek.com/blog/main/archives/2009/05/us_teens_sent_2.html.

13. Steven Cherry, "thx 4 the revnu," *IEEE Spectrum*, October 2008.

14. Sara Rimer, "Play with Your Food, Just Don't Text!" *New York Times*, May 26, 2009.

15. Nielsen Company, "A2/M2 Three Screen Report: 1st Quarter 2009," May 20, 2009, http://blog.nielsen.com/nielsenwire/wp-content/uploads/2009/05/nielsen_threescreenreport_q109.pdf.

16. Forrester Research, "How European Teens Consume Media," December 4, 2009, www.forrester.com/rb/Research/how_european_teens_consume_media/q/id/53763/t/2.

17. Heidi Dawley, "Time-wise, Internet Is Now TV's Equal," *Media Life*, February 1, 2006.

18. Council for Research Excellence, "The Video Consumer Mapping Study," March 26, 2009, www.researchexcellence.com/vcm_overview.pdf.

19. Bureau of Labor Statistics, "American Time Use Survey," 2004–2008, www .bls.gov/tus/.

20. Noreen O'Leary, "Welcome to My World," *Adweek*, November 17, 2008.

21. Marshall McLuhan, *Understanding Media: The Extensions of Man*, critical ed., ed. W. Terrence Gordon (Corte Madera, CA: Gingko, 2003), 237.

22. Anne Mangen, "Hypertext Fiction Reading: Haptics and Immersion," *Journal of Research in Reading*, 31, no. 4 (2008): 404–19.

23. Cory Doctorow, "Writing in the Age of Distraction," *Locus*, January 2009.

24. Ben Sisario, "Music Sales Fell in 2008, but Climbed on the Web," *New York Times*, December 31, 2008.

25. Ronald Grover, "Hollywood Is Worried as DVD Sales Slow," *BusinessWeek*, February 19, 2009; Richard Corliss, "Why Netflix Stinks," *Time*, August 10, 2009.

26. Chrystal Szeto, "U.S. Greeting Cards and Postcards," Pitney Bowes Background Paper No. 20, November 21, 2005, www.postinsight.com/files/ Nov21_GreetingCards_Final.pdf.

27. Brigid Schulte, "So Long, Snail Shells," *Washington Post*, July 25, 2009.

28. Scott Jaschik, "Farewell to the Printed Monograph," *Inside Higher Ed*, March 23, 2009, www.insidehighered.com/news/2009/03/23/Michigan.

29. Arnold Schwarzenegger, "Digital Textbooks Can Save Money, Improve Learning," *Mercury News*, June 7, 2009.

30. Tim Arango, "Fall in Newspaper Sales Accelerates to Pass 7%," *New York Times*, April 27, 2009.

31. David Cook, "Monitor Shifts from Print to Web-Based Strategy," *Christian Science Monitor*, October 28, 2008.

32. Tom Hall, "'We Will Never Launch Another Paper,'" *PrintWeek*, February 20, 2009, www.printweek.com/news/881913/We-will-launch-paper.

33. Tyler Cowen, *Create Your Own Economy* (New York: Dutton, 2009), 43.

34. Michael Scherer, "Does Size Matter?," *Columbia Journalism Review*, November/December 2002.

35. Quoted in Carl R. Ramey, *Mass Media Unleashed* (Lanham, MD: Rowman & Littlefield, 2007), 123.

36. Jack Shafer, "The *Times'* New Welcome Mat," *Slate*, April 1, 2008, www .slate.com/id/2187884.

37. Kathleen Deveny, "Reinventing Newsweek," *Newsweek*, May 18, 2009.

38. Carl DiOrio, "Warners Teams with Facebook for 'Watchmen,'" *Hollywood Reporter*, May 11, 2009, www.hollywoodreporter.com/hr/content_display/news/e3i4b5caa365ad73b3a32b7e20ib5eae9c0.

39. Sarah McBride, "The Way We'll Watch," *Wall Street Journal*, December 8, 2008.

40. Dave Itzkoff, "A Different Tweet in Beethoven's 'Pastoral,'" *New York Times*, July 24, 2009.

41. Stephanie Clifford, "Texting at a Symphony? Yes, but Only to Select an Encore," *New York Times*, May 15, 2009.

42. The nine hundred–member Westwinds Community Church, in Jackson, Michigan, has been a pacesetter in weaving social networking into services. During sermons, congregants send messages through Twitter, and the tweets unspool on large video screens. One message sent during a 2009 service read, according to a report in *Time* magazine, "I have a hard time recognizing God in the middle of everything." Bonnie Rochman, "Twittering in Church," *Time*, June 1, 2009.

43. Chrystia Freeland, "View from the Top: Eric Schmidt of Google," *Financial Times*, May 21, 2009.

44. John Carlo Bertot, Charles R. McClure, Carla B. Wright, et al., "Public Libraries and the Internet 2008: Study Results and Findings," Information Institute of the Florida State University College of Information, 2008; American Library Association, "Libraries Connect Communities: Public Library Funding & Technology Access Study 2008–2009," September 25, 2009, www.ala.org/ala/research/initiatives/plftas/2008_2009/librariesconnectcommunities3.pdf.

45. Scott Corwin, Elisabeth Hartley, and Harry Hawkes, "The Library Rebooted," *Strategy & Business*, Spring 2009.

Six THE VERY IMAGE OF A BOOK

1. Ting-i Tsai and Geoffrey A. Fowler, "Race Heats Up to Supply E-Reader Screens," *Wall Street Journal*, December 29, 2009.

2. Motoko Rich, "Steal This Book (for $9.99)," *New York Times*, May 16, 2009; Brad Stone, "Best Buy and Verizon Jump into E-Reader Fray," *New York Times*, September 22, 2009; Brad Stone and Motoko Rich, "Turning Page, E-Books Start to Take Hold," *New York Times*, December 23, 2008.

3. Jacob Weisberg, "Curling Up with a Good Screen," *Newsweek*, March 30, 2009. The italics are Weisberg's.

4. Charles McGrath, "By-the-Book Reader Meets the Kindle," *New York Times*, May 29, 2009.

5. L. Gordon Crovitz, "The Digital Future of Books," *Wall Street Journal*, May 19, 2008.

6. Debbie Stier, "Are We Having the Wrong Conversation about EBook Pricing?," HarperStudio blog, February 26, 2009, http://theharperstudio .com/2009/02/are-we-having-the-wrong-conversation-about-ebook-pricing.

7. Steven Johnson, "How the E-Book Will Change the Way We Read and Write," *Wall Street Journal*, April 20, 2009.

8. Christine Rosen, "People of the Screen," *New Atlantis*, Fall 2008.

9. David A. Bell, "The Bookless Future: What the Internet Is Doing to Scholarship," *New Republic*, May 2, 2005.

10. John Updike, "The End of Authorship," *New York Times Sunday Book Review*, June 25, 2006.

11. Norimitsu Onishi, "Thumbs Race as Japan's Best Sellers Go Cellular," *New York Times*, January 20, 2008. See also Dana Goodyear, "I ♥ Novels," *New Yorker*, December 22, 2008.

12. Tim O'Reilly, "Reinventing the Book in the Age of the Web," *O'Reilly Radar* blog, April 29, 2009, http://radar.oreilly.com/2009/04/reinventing-the-book-age-of-web.html.

13. Motoko Rich, "Curling Up with Hybrid Books, Videos Included," *New York Times*, September 30, 2009.

14. Johnson, "How the E-Book Will Change."

15. Andrew Richard Albanese, "Q&A: The Social Life of Books," *Library Journal*, May 15, 2006.

16. Kevin Kelly, "Scan this Book!" *New York Times Magazine*, May 14, 2006.

17. Caleb Crain, "How Is the Internet Changing Literary Style?," *Steamboats Are Ruining Everything* blog, June 17, 2008, www.steamthing.com/2008/06/how-is-the-inte.html.

18. Some Kindle owners received a startling lesson in the ephemerality of digital text when, on the morning of July 17, 2009, they awoke to find that the e-book versions of George Orwell's *1984* and *Animal Farm* they had purchased from Amazon.com had disappeared from their devices. It turned out that Amazon had erased the books from customers' Kindles after discovering that the editions were unauthorized.

19. Up to now, concerns about the influence of digital media on language have centered on the abbreviations and emoticons that kids use in instant messaging and texting. But such affectations will probably prove benign, just the latest twist in the long history of slang. Adults would be wiser to pay

attention to how their own facility with writing is changing. Is their vocab-ulary shrinking or becoming more hackneyed? Is their syntax becoming less flexible and more formulaic? Those are the types of questions that matter in judging the Net's long-run effects on the range and expressive-ness of language.

20. Wendy Griswold, Terry McDonnell, and Nathan Wright, "Reading and the Reading Class in the Twenty-First Century," *Annual Review of Sociology*, 31 (2005): 127–41. See also Caleb Crain, "Twilight of the Books," *New Yorker*, December 24, 2007.

21. Steven Levy, "The Future of Reading," *Newsweek*, November 26, 2007.

22. Alphonse de Lamartine, *Ouvres Diverses* (Brussels: Louis Hauman, 1836), 106–7. Translation by the author.

23. Philip G. Hubert, "The New Talking Machines," *Atlantic Monthly*, February 1889.

24. Edward Bellamy, "With the Eyes Shut," *Harper's*, October 1889.

25. Octave Uzanne, "The End of Books," *Scribner's Magazine*, August 1894.

26. George Steiner, "Ex Libris," *New Yorker*, March 17, 1997.

27. Mark Federman, "Why Johnny and Janey Can't Read, and Why Mr. and Mrs. Smith Can't Teach: The Challenge of Multiple Media Literacies in a Tumultuous Time," undated, http://individual.utoronto.ca/markfederman /WhyJohnnyandJaneyCantRead.pdf.

28. Clay Shirky, "Why Abundance Is Good: A Reply to Nick Carr," *Encyclopae-dia Britannica Blog*, July 17, 2008, www.britannica.com/blogs/2008/07/ why-abundance-is-good-a-reply-to-nick-carr.

29. Alberto Manguel, *The Library at Night* (New Haven, CT: Yale University Press, 2008), 218.

30. David M. Levy, *Scrolling Forward: Making Sense of Documents in the Digital Age* (New York: Arcade, 2001), 101–2.

Seven THE JUGGLER'S BRAIN

1. Katie Hafner, "Texting May Be Taking a Toll," *New York Times*, May 25, 2009.

2. Torkel Klingberg, *The Overflowing Brain: Information Overload and the Limits of Working Memory*, trans. Neil Betteridge (Oxford: Oxford Univer-sity Press, 2009), 166–67.

3. Ap Dijksterhuis, "Think Different: The Merits of Unconscious Thought in Preference Development and Decision Making," *Journal of Personality and Social Psychology*, 87, no. 5 (2004): 586–98.

4. Marten W. Bos, Ap Dijksterhuis, and Rick B. van Baaren, "On the Goal-Dependency of Unconscious Thought," *Journal of Experimental Social Psychology*, 44 (2008): 1114–20.

5. Stefanie Olsen, "Are We Getting Smarter or Dumber?," CNET News, September 21, 2005, http://news.cnet.com/Are-we-getting-smarter-or-dumber/2008-1008_3-5875404.html.

6. Michael Merzenich, "Going Googly," *On the Brain* blog, August 11, 2008, http://merzenich.positscience.com/?p=177.

7. Gary Small and Gigi Vorgan, *iBrain: Surviving the Technological Alteration of the Modern Mind* (New York: Collins, 2008), 1.

8. G. W. Small, T. D. Moody, P. Siddarth, and S. Y. Bookheimer, "Your Brain on Google: Patterns of Cerebral Activation during Internet Searching," *American Journal of Geriatric Psychiatry*, 17, no. 2 (February 2009): 116–26. See also Rachel Champeau, "UCLA Study Finds That Searching the Internet Increases Brain Function," UCLA Newsroom, October 14, 2008, http://newsroom.ucla.edu/portal/ucla/ucla-study-finds-that-searching-64348.aspx.

9. Small and Vorgan, *iBrain*, 16–17.

10. Maryanne Wolf, interview with the author, March 28, 2008.

11. Steven Johnson, *Everything Bad Is Good for You: How Today's Popular Culture Is Actually Making Us Smarter* (New York: Riverhead Books, 2005), 19.

12. John Sweller, *Instructional Design in Technical Areas* (Camberwell, Australia: Australian Council for Educational Research, 1999), 4.

13. Ibid., 7.

14. Ibid.

15. Ibid., 11.

16. Ibid., 4–5. For a broad review of current thinking on the limits of working memory, see Nelson Cowan, *Working Memory Capacity* (New York: Psychology Press, 2005).

17. Klingberg, *Overflowing Brain*, 39 and 72–75.

18. Sweller, *Instructional Design*, 22.

19. George Landow and Paul Delany, "Hypertext, Hypermedia and Literary Studies: The State of the Art," in *Multimedia: From Wagner to Virtual Reality*, ed. Randall Packer and Ken Jordan (New York: Norton, 2001), 206–16.

20. Jean-Francois Rouet and Jarmo J. Levonen, "Studying and Learning with Hypertext: Empirical Studies and Their Implications," in *Hypertext and Cognition*, ed. Jean-Francois Rouet, Jarmo J. Levonen, Andrew Dillon, and Rand J. Spiro (Mahwah, NJ: Erlbaum, 1996), 16–20.

21. David S. Miall and Teresa Dobson, "Reading Hypertext and the Experience of Literature," *Journal of Digital Information*, 2, no. 1 (August 13, 2001).

22. D. S. Niederhauser, R. E. Reynolds, D. J. Salmen, and P. Skolmoski, "The Influence of Cognitive Load on Learning from Hypertext," *Journal of Educational Computing Research*, 23, no. 3 (2000): 237–55.

23. Erping Zhu, "Hypermedia Interface Design: The Effects of Number of Links and Granularity of Nodes," *Journal of Educational Multimedia and Hypermedia*, 8, no. 3 (1999): 331–58.

24. Diana DeStefano and Jo-Anne LeFevre, "Cognitive Load in Hypertext Reading: A Review," *Computers in Human Behavior*, 23, no. 3 (May 2007): 1616–41. The paper was originally published online on September 30, 2005.

25. Steven C. Rockwell and Loy A. Singleton, "The Effect of the Modality of Presentation of Streaming Multimedia on Information Acquisition," *Media Psychology*, 9 (2007): 179–91.

26. Helene Hembrooke and Geri Gay, "The Laptop and the Lecture: The Effects of Multitasking in Learning Environments," *Journal of Computing in Higher Education*, 15, no. 1 (September 2003): 46–64.

27. Lori Bergen, Tom Grimes, and Deborah Potter, "How Attention Partitions Itself during Simultaneous Message Presentations," *Human Communication Research*, 31, no. 3 (July 2005): 311–36.

28. Sweller, *Instructional Design*, 137–47.

29. K. Renaud, J. Ramsay, and M. Hair, " 'You've Got Email!' Shall I Deal with It Now?," *International Journal of Human-Computer Interaction*, 21, no. 3 (2006): 313–32.

30. See, for example, J. Gregory Trafton and Christopher A. Monk, "Task Interruptions," *Reviews of Human Factors and Ergonomics*, 3 (2008): 111–26. Researchers believe that frequent interruptions lead to cognitive overload and impair the formation of memories.

31. Maggie Jackson, *Distracted: The Erosion of Attention and the Coming Dark Age* (Amherst, NY: Prometheus, 2008), 79.

32. Karin Foerde, Barbara J. Knowlton, and Russell A. Poldrack, "Modulation of Competing Memory Systems by Distraction," *Proceedings of the National Academy of Sciences*, 103, no. 31 (August 1, 2006): 11778–83; and "Multi-Tasking Adversely Affects Brain's Learning," University of California press release, July 7, 2005.

33. Christopher F. Chabris, "You Have Too Much Mail," *Wall Street Journal*, December 15, 2008. The italics are Chabris's.

34. Sav Shrestha and Kelsi Lenz, "Eye Gaze Patterns While Searching vs. Browsing a Website," *Usability News*, 9, no. 1 (January 2007), www.surl.org/usabilitynews/91/eyegaze.asp.

35. Jakob Nielsen, "F-Shaped Pattern for Reading Web Content," *Alertbox*, April 17, 2006, www.useit.com/alertbox/reading_pattern.html.

36. Jakob Nielsen, "How Little Do Users Read?," *Alertbox*, May 6, 2008, www .useit.com/alertbox/percent-text-read.html.

37. Harald Weinreich, Hartmut Obendorf, Eelco Herder, and Matthias Mayer, "Not Quite the Average: An Empirical Study of Web Use," *ACM Transactions on the Web*, 2, no. 1 (2008).

38. Jakob Nielsen, "How Users Read on the Web," *Alertbox*, October 1, 1997, www.useit.com/alertbox/9710a.html.

39. "Puzzling Web Habits across the Globe," *ClickTale* blog, July 31, 2008, www .clicktale.com/2008/07/31/puzzling-web-habits-across-the-globe-part-1/.

40. University College London, "Information Behaviour of the Researcher of the Future," January 11, 2008, www.ucl.ac.uk/slais/research/ciber/down loads/ggexecutive.pdf.

41. Merzenich, "Going Googly."

42. Ziming Liu, "Reading Behavior in the Digital Environment," *Journal of Documentation*, 61, no. 6 (2005): 700–712.

43. Shawn Green and Daphne Bavelier, "Action Video Game Modifies Visual Selective Attention," *Nature*, 423 (May 29, 2003): 534–37.

44. Elizabeth Sillence, Pam Briggs, Peter Richard Harris, and Lesley Fishwick, "How Do Patients Evaluate and Make Use of Online Health Information?," *Social Science and Medicine*, 64, no. 9 (May 2007): 1853–62.

45. Klingberg, *Overflowing Brain*, 115–24.

46. Small and Vorgan, *iBrain*, 21.

47. Sam Anderson, "In Defense of Distraction," *New York*, May 25, 2009.

48. Quoted in Don Tapscott, *Grown Up Digital* (New York: McGraw-Hill, 2009), 108–9.

49. Quoted in Jackson, *Distracted*, 79–80.

50. Quoted in Sharon Begley and Janeen Interlandi, "The Dumbest Generation? Don't Be Dumb," *Newsweek*, June 2, 2008.

51. Lucius Annaeus Seneca, *Letters from a Stoic* (New York: Penguin Classics, 1969), 33.

52. Patricia M. Greenfield, "Technology and Informal Education: What Is Taught, What Is Learned," *Science*, 323, no. 5910 (January 2, 2009): 69–71.

53. Eyal Ophir, Clifford Nass, and Anthony D. Wagner, "Cognitive Control in Media Multitaskers," *Proceedings of the National Academy of Sciences*, August 24, 2009, www.pnas.org/content/early/2009/08/21/0903620106.full .pdf. See also Adam Gorlick, "Media Multitaskers Pay Mental Price, Stanford

Study Shows," *Stanford Report*, August 24, 2009, http://news.stanford.edu/news/2009/august24/multitask-research-study-082409.html.

54. Michael Merzenich, interview with the author, September 11, 2009.

55. James Boswell, *The Life of Samuel Johnson, LL. D.* (London: Bell, 1889), 331–32.

a digression   ON THE BUOYANCY OF IQ SCORES

1. Don Tapscott, *Grown Up Digital* (New York: McGraw-Hill, 2009), 291.

2. College Board, "PSAT/NMSQT Data & Reports," http://professionals.collegeboard.com/data-reports-research/psat.

3. Naomi S. Baron, *Always On: Language in an Online and Mobile World* (Oxford: Oxford University Press, 2008), 202.

4. David Schneider, "Smart as We Can Get?," *American Scientist*, July–August 2006.

5. James R. Flynn, "Requiem for Nutrition as the Cause of IQ Gains: Raven's Gains in Britain 1938–2008," *Economics and Human Biology*, 7, no. 1 (March 2009): 18–27.

6. Some contemporary readers may find Flynn's choice of words insensitive. He explains, "We are in a transitional period in which the term 'mentally retarded' is being replaced by the term 'mentally disabled' in the hope of finding words with a less negative connotation. I have retained the old term for clarity and because history has shown that negative connotations are simply passed on from one label to another." James R. Flynn, *What Is Intelligence? Beyond the Flynn Effect* (Cambridge: Cambridge University Press, 2007), 9–10.

7. Ibid., 9.

8. Ibid., 172–73.

9. "The World Is Getting Smarter," *Intelligent Life*, December 2007. See also Matt Nipert, "Eureka!" *New Zealand Listener*, October 6–12, 2007.

10. Patricia M. Greenfield, "Technology and Informal Education: What Is Taught, What Is Learned," *Science*, 323, no. 5910 (January 2, 2009): 69–71.

11. Denise Gellene, "IQs Rise, but Are We Brighter?," *Los Angeles Times*, October 27, 2007.

Eight   THE CHURCH OF GOOGLE

1. For an account of Taylor's life, see Robert Kanigel, *One Best Way: Frederick Winslow Taylor and the Enigma of Efficiency* (New York: Viking, 1997).

2. Frederick Winslow Taylor, *The Principles of Scientific Management* (New York: Harper, 1911), 25.

3. Ibid., 7.

4. Google Inc. Press Day Webcast, May 10, 2006, http://google.client.share holder.com/Visitors/event/build2/MediaPresentation.cfm?MediaID= 20263&Player=1.

5. Marissa Mayer, "Google I/O '08 Keynote," YouTube, June 5, 2008, www .youtube.com/watch?v=6xocAzQ7PVs.

6. Bala Iyer and Thomas H. Davenport, "Reverse Engineering Google's Innovation Machine," *Harvard Business Review*, April 2008.

7. Anne Aula and Kerry Rodden, "Eye-Tracking Studies: More than Meets the Eye," *Official Google Blog*, February 6, 2009, http://googleblog.blogspot .com/2009/02/eye-tracking-studies-more-than-meets.html.

8. Helen Walters, "Google's Irene Au: On Design Challenges," *BusinessWeek*, March 18, 2009.

9. Mayer, "Google I/O '08 Keynote."

10. Laura M. Holson, "Putting a Bolder Face on Google," *New York Times*, February 28, 2009.

11. Neil Postman, *Technopoly: The Surrender of Culture to Technology* (New York: Vintage, 1993), 51.

12. Ken Auletta, *Googled: The End of the World as We Know It* (New York: Penguin, 2009), 22.

13. Google, "Company Overview," undated, www.google.com/corporate.

14. Kevin J. Delaney and Brooks Barnes, "For Soaring Google, Next Act Won't Be So Easy," *Wall Street Journal*, June 30, 2005.

15. Google, "Technology Overview," undated, www.google.com/corporate/ tech.html.

16. Academy of Achievement, "Interview: Larry Page," October 28, 2000, www.achievement.org/autodoc/page/pag0int-1.

17. John Battelle, *The Search: How Google and Its Rivals Rewrote the Rules of Business and Transformed Our Culture* (New York: Portfolio, 2005), 66–67.

18. Ibid.

19. See Google, "Google Milestones," undated, www.google.com/corporate/ history.html.

20. Sergey Brin and Lawrence Page, "The Anatomy of a Large-Scale Hypertextual Web Search Engine," *Computer Networks*, 30 (April 1, 1998): 107–17.

21. Walters, "Google's Irene Au."

22. Mark Zuckerberg, "Improving Your Ability to Share and Connect,"

Facebook blog, March 4, 2009, http://blog.facebook.com/blog.php?post =57822962130.

23. Saul Hansell, "Google Keeps Tweaking Its Search Engine," *New York Times*, June 3, 2007.

24. Brennon Slattery, "Google Caffeinates Its Search Engine," *PC World*, August 11, 2009, www.pcworld.com/article/169989.

25. Nicholas Carlson, "Google Co-Founder Larry Page Has Twitter-Envy," *Silicon Alley Insider*, May 19, 2009, www.businessinsider.com/google-cofounder -larry-page-has-twitter-envy-2009-5.

26. Kit Eaton, "Developers Start to Surf Google Wave, and Love It," *Fast Company*, July 21, 2009, www.fastcompany.com/blog/kit-eaton/technomix/ developers-start-surf-google-wave-and-love-it.

27. Doug Caverly, "New Report Slashes YouTube Loss Estimate by $300M," *WebProNews*, June 17, 2009, www.webpronews.com/topnews/2009/06/17/ new-report-slashes-youtube-loss-estimate-by-300m.

28. Richard MacManus, "Store 100%—Google's Golden Copy," *ReadWriteWeb*, March 5, 2006, www.readwriteweb.com/archives/store_100_googl.php.

29. Jeffrey Toobin, "Google's Moon Shot," *New Yorker*, February 5, 2007.

30. Jen Grant, "Judging Book Search by Its Cover," *Official Google Blog*, November 17, 2005, http://googleblog.blogspot.com/2005/11/judging-book-search-by-its-cover.html.

31. See U.S. Patent no. 7,508,978.

32. Google, "History of Google Books," undated, http://books.google.com/ googlebooks/history.html.

33. Authors Guild, "Authors Guild Sues Google, Citing 'Massive Copyright Infringement,'" press release, September 20, 2005.

34. Eric Schmidt, "Books of Revelation," *Wall Street Journal*, October 18, 2005.

35. U.S. District Court, Southern District of New York, "Settlement Agreement: The Authors Guild, Inc., Association of American Publishers, Inc., et al., Plaintiffs, v. Google Inc., Defendant," Case No. 05 CV 8136-JES, October 28, 2008.

36. American Library Association, "Library Association Comments on the Proposed Settlement," filing with the U.S. District Court, Southern District of New York, Case No. 05 CV 8136-DC, May 4, 2009.

37. Robert Darnton, "Google and the Future of Books," *New York Review of Books*, February 12, 2009.

38. Richard Koman, "Google, Books and the Nature of Evil," *ZDNet Government* blog, April 30, 2009, http://government.zdnet.com/?p=4725.

39. In what may be a harbinger of the future, a prestigious Massachusetts prep school, Cushing Academy, announced in 2009 that it was removing all the books from its library and replacing them with desktop computers, flat-screen TVs, and a score of Kindles and other e-readers. The school's headmaster, James Tracy, proclaimed the bookless library "a model for the 21st-century school." David Abel, "Welcome to the Library. Say Goodbye to the Books," *Boston Globe*, September 4, 2009.

40. Alexandra Alter, "The Next Age of Discovery," *Wall Street Journal*, May 8, 2009.

41. Adam Mathes, "Collect, Share, and Discover Books," *Official Google Blog*, September 6, 2007, http://googleblog.blogspot.com/2007/09/collect-share-and-discover-books.html.

42. Manas Tungare, "Share and Enjoy," *Inside Google Books* blog, September 6, 2007, http://booksearch.blogspot.com/2007/08/share-and-enjoy.html.

43. Bill Schilit and Okan Kolak, "Dive into the Meme Pool with Google Book Search," *Inside Google Books* blog, September 6, 2007, http://booksearch.blogspot.com/2007/09/dive-into-meme-pool-with-google-book.html; and Diego Puppin, "Explore a Book in 10 Seconds," *Inside Google Books* blog, July 1, 2009, http://booksearch.blogspot.com/2009/06/explore-book-in-10-seconds.html.

44. Passages from Hawthorne's notebooks are quoted in Julian Hawthorne, *Nathaniel Hawthorne and His Wife: A Biography*, vol. 1 (Boston: James R. Osgood, 1885), 498–503.

45. Leo Marx, *The Machine in the Garden: Technology and the Pastoral Ideal in America* (New York: Oxford University Press, 2000), 28–29.

46. Quoted in Will Durant and Ariel Durant, *The Age of Reason Begins* (New York: Simon & Schuster, 1961), 65.

47. Vannevar Bush, "As We May Think," *Atlantic Monthly*, July 1945.

48. David M. Levy, "To Grow in Wisdom: Vannevar Bush, Information Overload, and the Life of Leisure," *Proceedings of the 5th ACM/IEEE-CS Joint Conference on Digital Libraries*, 2005, 281–86.

49. Ibid.

50. Ralph Waldo Emerson, "Books," *Atlantic Monthly*, January 1858.

51. Larry Page, keynote address before AAAS Annual Conference, San Francisco, February 16, 2007, http://news.cnet.com/1606-2_3-6160334.html.

52. Academy of Achievement, "Interview: Larry Page."

53. Rachael Hanley, "From Googol to Google: Co-founder Returns," *Stanford Daily*, February 12, 2003.

54. Academy of Achievement, "Interview: Larry Page."

55. Steven Levy, "All Eyes on Google," *Newsweek*, April 12, 2004.

56. Spencer Michaels, "The Search Engine That Could," *NewsHour with Jim Lehrer*, November 29, 2002.

57. See Richard MacManus, "Full Text of Google Analyst Day Powerpoint Notes," *Web 2.0 Explorer* blog, March 7, 2006, http://blogs.zdnet.com/web2explorer/?p=132.

58. Quoted in Jean-Pierre Dupuy, *On the Origins of Cognitive Science: The Mechanization of the Mind* (Cambridge, MA: MIT Press, 2009), xiv.

59. George B. Dyson, *Darwin among the Machines: The Evolution of Global Intelligence* (Reading, MA: Addison-Wesley, 1997), 10.

60. George Dyson, "Turing's Cathedral," *Edge*, October 24, 2005, www.edge.org/3rd_culture/dyson05/dyson_05index.html.

61. Greg Jarboe, "A 'Fireside Chat' with Google's Sergey Brin," *Search Engine Watch*, October 16, 2003, http://searchenginewatch.com/3081081.

62. See Pamela McCorduck, *Machines Who Think: A Personal Inquiry into the History and Prospects of Artificial Intelligence* (Natick, MA: Peters, 2004), 111.

63. Lewis Mumford, *The Myth of the Machine: Technics and Human Development* (New York: Harcourt Brace Jovanovitch, 1967), 29.

64. David G. Stork, ed., *HAL's Legacy: 2001's Computer as Dream and Reality* (Cambridge, MA: MIT Press, 1996), 165-66.

65. John von Neumann, *The Computer and the Brain*, 2nd ed. (New Haven, CT: Yale University Press, 2000), 82. The italics are von Neumann's.

66. Ari N. Schulman, "Why Minds Are Not like Computers," *New Atlantis*, Winter 2009.

Nine SEARCH, MEMORY

1. Quoted in Alberto Manguel, *A History of Reading* (New York: Viking, 1996), 49.

2. Umberto Eco, "From Internet to Gutenberg," lecture presented at Columbia University's Italian Academy for Advanced Studies in America, November 12, 1996, www.umbertoeco.com/en/from-internet-to-gutenberg-1996.html.

3. Quoted in Ann Moss, *Printed Commonplace-Books and the Structuring of Renaissance Thought* (Oxford: Oxford University Press, 1996), 102-4.

4. Erika Rummel, "Erasmus, Desiderius," in *Philosophy of Education*, ed. J. J. Chambliss (New York: Garland, 1996), 198.

5. Quoted in Moss, *Printed Commonplace-Books*, 12.

6. Ann Moss writes that "the commonplace-book was part of the initial intellectual experience of every schoolboy" in the Renaissance. *Printed Commonplace-Books*, viii.

7. Francis Bacon, *The Works of Francis Bacon*, vol. 4, ed. James Spedding, Robert Leslie Ellis, and Douglas Denon Heath (London: Longman, 1858), 435.

8. Naomi S. Baron, *Always On: Language in an Online and Mobile World* (Oxford: Oxford University Press, 2008), 197.

9. Clive Thompson, "Your Outboard Brain Knows All," *Wired*, October 2007.

10. David Brooks, "The Outsourced Brain," *New York Times*, October 26, 2007.

11. Peter Suderman, "Your Brain Is an Index," *American Scene*, May 10, 2009, www.theamericanscene.com/2009/05/11/your-brain-is-an-index.

12. Alexandra Frean, "Google Generation Has No Need for Rote Learning," *Times* (London), December 2, 2008; and Don Tapscott, *Grown Up Digital* (New York: McGraw-Hill, 2009), 115.

13. Saint Augustine, *Confessions*, trans. Henry Chadwick (New York: Oxford University Press, 1998), 187.

14. William James, *Talks to Teachers on Psychology: And to Students on Some of Life's Ideals* (New York: Holt, 1906), 143.

15. See Eric R. Kandel, *In Search of Memory: The Emergence of a New Science of Mind* (New York: Norton, 2006), 208–10.

16. Ibid., 210–11.

17. Louis B. Flexner, Josefa B. Flexner, and Richard B. Roberts, "Memory in Mice Analyzed with Antibiotics," *Science*, 155 (1967): 1377–83.

18. Kandel, *In Search of Memory*, 221.

19. Ibid., 214–15.

20. Ibid., 221.

21. Ibid., 276.

22. Ibid.

23. Ibid., 132.

24. Until his name was disclosed upon his death in 2008, Molaison was referred to in the scientific literature as H.M.

25. See Larry R. Squire and Pablo Alvarez, "Retrograde Amnesia and Memory Consolidation: A Neurobiological Perspective," *Current Opinion in Neurobiology*, 5 (1995): 169–77.

26. Daniel J. Siegel, *The Developing Mind* (New York: Guilford, 2001), 37–38.

27. In a 2009 study, French and American researchers found evidence that

brief, intense oscillations that ripple through the hippocampus during sleep play an important role in storing memories in the cortex. When the researchers suppressed the oscillations in the brains of rats, the rats were unable to consolidate long-term spatial memories. Gabrielle Girardeau, Karim Benchenane, Sidney I. Wiener, et al., "Selective Suppression of Hippocampal Ripples Impairs Spatial Memory," *Nature Neuroscience*, September 13, 2009, www.nature.com/neuro/journal/vaop/ncurrent/abs/nn.2384.html.

28. University of Haifa, "Researchers Identified a Protein Essential in Long Term Memory Consolidation," Physorg.com, September 9, 2008, www.physorg.com/news140173258.html.

29. See Jonah Lehrer, *Proust Was a Neuroscientist* (New York: Houghton Mifflin, 2007), 84–85.

30. Joseph LeDoux, *Synaptic Self: How Our Brains Become Who We Are* (New York: Penguin, 2002), 161.

31. Nelson Cowan, *Working Memory Capacity* (New York: Psychology Press, 2005), 1.

32. Torkel Klingberg, *The Overflowing Brain: Information Overload and the Limits of Working Memory*, trans. Neil Betteridge (Oxford: Oxford University Press, 2009), 36.

33. Sheila E. Crowell, "The Neurobiology of Declarative Memory," in John H. Schumann, Shelia E. Crowell, Nancy E. Jones, et al., *The Neurobiology of Learning: Perspectives from Second Language Acquisition* (Mahwah, NJ: Erlbaum, 2004), 76.

34. See, for example, Ray Hembree and Donald J. Dessart, "Effects of Hand-held Calculators in Precollege Mathematics Education: A Meta-analysis," *Journal for Research in Mathematics Education*, 17, no. 2 (1986): 83–99.

35. Kandel, *In Search of Memory*, 210.

36. Quoted in Maggie Jackson, *Distracted: The Erosion of Attention and the Coming Dark Age* (Amherst, NY: Prometheus, 2008), 242.

37. Kandel, *In Search of Memory*, 312–15.

38. David Foster Wallace, *This Is Water: Some Thoughts, Delivered on a Significant Occasion, about Living a Compassionate Life* (New York: Little, Brown, 2009), 54 and 123.

39. Ari N. Schulman, correspondence with the author, June 7, 2009.

40. Lea Winerman, "The Culture of Memory," *Monitor on Psychology*, 36, no. 8 (September 2005): 56.

41. Pascal Boyer and James V. Wertsch, eds., *Memory in Mind and Culture* (New York: Cambridge University Press, 2009), 7 and 288.

42. Richard Foreman, "The Pancake People, or, 'The Gods Are Pounding My Head,'" *Edge*, March 8, 2005, www.edge.org/3rd_culture/foreman05/fore man05_index.html.

a digression ON THE WRITING OF THIS BOOK

1. Benjamin Kunkel, "Lingering," *n+1*, May 31, 2009, www.nplusonemag .com/lingering. The italics are Kunkel's.

Ten A THING LIKE ME

1. Joseph Weizenbaum, "ELIZA—A Computer Program for the Study of Natural Language Communication between Man and Machine," *Communications of the Association for Computing Machinery*, 9, no. 1 (January 1966): 36–45.

2. David Golumbia, *The Cultural Logic of Computation* (Cambridge, MA: Harvard University Press, 2009), 42.

3. Quoted in Golumbia, *Cultural Logic*, 37.

4. Ibid., 42.

5. Weizenbaum, "ELIZA."

6. Ibid.

7. Joseph Weizenbaum, *Computer Power and Human Reason: From Judgment to Calculation* (New York: Freeman, 1976), 5.

8. Ibid., 189.

9. Ibid., 7.

10. Quoted in Weizenbaum, *Computer Power*, 5.

11. Kenneth Mark Colby, James B. Watt, and John P. Gilbert, "A Computer Method of Psychotherapy: Preliminary Communication," *Journal of Nervous and Mental Disease*, 142, no. 2 (1966): 148–52.

12. Weizenbaum, *Computer Power*, 8.

13. Ibid., 17–38.

14. Ibid., 227.

15. John McCarthy, "An Unreasonable Book," *SIGART Newsletter*, 58 (June 1976).

16. Michael Balter, "Tool Use Is Just Another Trick of the Mind," *Science-NOW*, January 28, 2008, http://sciencenow.sciencemag.org/cgi/content/full/2008/128/2.

17. *The Letters of T. S. Eliot*, vol. 1, *1898–1922*, ed. Valerie Eliot (New York:

Harcourt Brace Jovanovich, 1988), 144. As for Nietzsche, his affair with the Malling-Hansen Writing Ball turned out to be as brief as it was intense. Like many of the early adopters of new gadgets who would follow in his eager footsteps, he became frustrated with the typewriter's flaws. The writing ball, it turned out, was buggy. When the Mediterranean air grew humid with the arrival of spring, the keys started to jam and the ink began to run on the page. The contraption, Nietzsche wrote in a letter, "is as delicate as a little dog and causes a lot of trouble." Within months he had given up on the writing ball, trading the balky device for a secretary, the young poet Lou Salomé, who transcribed his words as he spoke them. Five years later, in one of his last books, *On the Genealogy of Morals*, Nietzsche made an eloquent argument against the mechanization of human thought and personality. He praised the contemplative state of mind through which we quietly and willfully "digest" our experiences. "The temporary shutting of the doors and windows of consciousness, the relief from the clamant alarums," he wrote, allows the brain "to make room again for the new, and above all for the more noble functions." Friedrich Nietzsche, *The Genealogy of Morals* (Mineola, NY: Dover, 2003), 34.

18. Norman Doidge, *The Brain That Changes Itself: Stories of Personal Triumph from the Frontiers of Brain Science* (New York: Penguin, 2007), 311.

19. John M. Culkin, "A Schoolman's Guide to Marshall McLuhan," *Saturday Review*, March 18, 1967.

20. Marshall McLuhan, *Understanding Media: The Extensions of Man*, critical ed., ed. W. Terrence Gordon (Corte Madera, CA: Gingko Press, 2003), 63–70.

21. Lewis Mumford, *Technics and Civilization* (New York: Harcourt Brace, 1963), 15.

22. Weizenbaum, *Computer Power*, 25.

23. Roger Dobson, "Taxi Drivers' Knowledge Helps Their Brains Grow," *Independent*, December 17, 2006.

24. Doidge, *Brain That Changes Itself*, 310–11.

25. Jason P. Mitchell, "Watching Minds Interact," in *What's Next: Dispatches on the Future of Science*, ed. Max Brockman (New York: Vintage, 2009), 78–88.

26. Bill Thompson, "Between a Rock and an Interface," *BBC News*, October 7, 2008, http://news.bbc.co.uk/2/hi/technology/7656843.stm.

27. Christof van Nimwegen, "The Paradox of the Guided User: Assistance Can Be Counter-effective," SIKS Dissertation Series No. 2008-09, Utrecht University, March 31, 2008. See also Christof van Nimwegen and Herre van

Oostendorp, "The Questionable Impact of an Assisting Interface on Performance in Transfer Situations," *International Journal of Industrial Ergonomics*, 39, no. 3 (May 2009): 501–8.

28. Ibid.

29. Ibid.

30. "Features: Query Suggestions," Google Web Search Help, undated, http://labs.google.com/suggestfaq.html.

31. James A. Evans, "Electronic Publication and the Narrowing of Science and Scholarship," *Science*, 321 (July 18, 2008): 395–99.

32. Ibid.

33. Thomas Lord, "Tom Lord on Ritual, Knowledge and the Web," *Rough Type* blog, November 9, 2008, www.roughtype.com/archives/2008/11/tom_lord_on_rit.php.

34. Marc G. Berman, John Jonides, and Stephen Kaplan, "The Cognitive Benefits of Interacting with Nature," *Psychological Science*, 19, no. 12 (December 2008): 1207–12.

35. Carl Marziali, "Nobler Instincts Take Time," USC Web site, April 14, 2009, http://college.usc.edu/news/stories/547/nobler-instincts-take-time.

36. Mary Helen Immordino-Yang, Andrea McColl, Hanna Damasio, and Antonio Damasio, "Neural Correlates of Admiration and Compassion," *Proceedings of the National Academy of Sciences*, 106, no. 19 (May 12, 2009): 8021–26.

37. Marziali, "Nobler Instincts."

38. L. Gordon Crovitz, "Information Overload? Relax," *Wall Street Journal*, July 6, 2009.

39. Sam Anderson, "In Defense of Distraction," *New York*, May 25, 2009.

40. Tyler Cowen, *Create Your Own Economy* (New York: Dutton, 2009), 10.

41. Jamais Cascio, "Get Smarter," *Atlantic*, July/August 2009.

42. Martin Heidegger, *Discourse on Thinking* (New York: Harper & Row, 1966), 56. The italics are Heidegger's.

43. Martin Heidegger, *The Question Concerning Technology and Other Essays* (New York: Harper & Row, 1977), 35.

Epilogue   HUMAN ELEMENTS

1. William Stewart, "Essays to Be Marked by 'Robots,'" *Times Education Supplement*, September 25, 2009.

Afterword   THE MOST INTERESTING THING IN THE WORLD

1. Nielsen Company, "The Nielsen Total Audience Report: Q1 2019" and "The Total Audience Report: December 2014."

2. U.S. Department of Labor, "American Time Use Survey—2018 Results," Bureau of Labor Statistics news release, June 19, 2019.

3. When people do pick up a book these days, it's usually still a printed one. The rapid rise in e-book sales that followed the 2007 introduction of the Amazon Kindle, which I described in Chapter 6, hit a wall in 2013. Since then, publishers have seen their e-book sales erode. Electronic books now account for twenty percent of overall book sales in the United States, down from a high of thirty percent. Once again, reports of the death of the printed book have proven premature.

4. Sally Andrews, David A. Ellis, Heather Shaw, and Lukasz Piwek, "Beyond Self-Report: Tools to Compare Estimated and Real-World Smartphone Use," *PLOS ONE,* October 28, 2015, https://doi.org/10.1371/journal .pone.0139004.

5. Ivan Krstic, "How iOS Security Really Works," presentation at Apple Worldwide Developers Conference (2016), https://developer.apple.com/videos/ play/wwdc2016/705/.

6. Cary Stothart, Ainsley Mitchum, and Courtney Yehnert, "The Attentional Cost of Receiving a Cell Phone Notification," *Journal of Experimental Psychology: Human Perception and Performance,* 41, no. 4 (2015): 893–97.

7. Russell B. Clayton, Glenn Leshner, and Anthony Almond, "The Extended iSelf: The Impact of iPhone Separation on Cognition, Emotion, and Physiology," *Journal of Computer-Mediated Communication,* 20, no. 2 (March 2015): 119–35.

8. Kostadin Kushlev, Jason Proulx, and Elizabeth W. Dunn, "'Silence Your Phones': Smartphone Notifications Increase Inattention and Hyperactivity Symptoms," *Proceedings of the 2016 CHI Conference on Human Factors in Computing Systems* (May 2016): 1011–20.

9. Adrian F. Ward, Kristen Duke, Ayelet Gneezy, and Maarten W. Bos, "Brain Drain: The Mere Presence of One's Own Smartphone Reduces Available Cognitive Capacity," *Journal of the Association for Consumer Research,* 2, no. 2 (April 2017): 140–54.

10. Bill Thornton, Alyson Faires, Maija Robbins, and Eric Rollins, "The Mere Presence of a Cell Phone May Be Distracting: Implications for Attention and Task Performance," *Social Psychology,* 45, no. 6 (2014): 479–88. Also see Clarissa T. Tanil and Min Hooi Yong, "Mobile Phones: The Effect of Its

Presence on Learning and Memory," bioRxiv preprint (2019), https://www
.biorxiv.org/content/10.1101/678094v1.

11. Seungyeon Lee, Myeong W. Kim, Ian M. McDonough, et al., "The Effects of
Cell Phone Use and Emotion-Regulation Style on College Students' Learning," *Applied Cognitive Psychology*, 31, no. 3 (May/June 2017): 360–66.

12. Louis-Philippe Beland and Richard Murphy, "Ill Communication: Technology, Distraction, and Student Performance," *Labour Economics*, 41 (August
2016): 61–76.

13. Andrew K. Przybylski and Netta Weinstein, "Can You Connect with Me
Now? How the Presence of Mobile Communication Technology Influ-
ences Face-to-Face Conversation Quality," *Journal of Social and Personal
Relationships*, 30, no. 3 (2012): 237–46.

14. Shalini Misra, Lulu Cheng, Jamie Genevie, and Miao Yuan, "The iPhone
Effect: The Quality of In-Person Social Interactions in the Presence of
Mobile Devices," *Environment and Behavior*, 48, no. 2 (2016): 275–98.
Another study found that strangers are considerably less likely to exchange
smiles when phones are around: Kostadin Kushlev, John F. Hunter, Jason
Proulx, et al., "Smartphones Reduce Smiles Between Strangers," *Comput-
ers in Human Behavior*, 91 (February 2019): 12–16.

15. Vinod Menon, "Salience Network," *Brain Mapping: An Encyclopedic Reference*, vol. 2 (Oxford: Elsevier, 2015), 597–611.

16. Ibid. See also Lucina Q. Uddin, *Salience Network of the Human Brain*
(Oxford: Elsevier, 2017).

17. See, for example, Adam Alter, *Irresistible: The Rise of Addictive Technology
and the Business of Keeping Us Hooked* (New York: Penguin, 2017).

18. Adrian F. Ward, "Supernormal: How the Internet Is Changing Our Memories and Our Minds," *Psychological Inquiry*, 24, no. 4 (2013): 341–48.

19. See, for example, Mattha Busby, "Social Media Copies Gambling Methods
'to Create Psychological Cravings,'" *The Guardian*, May 8, 2018.

20. Erica Pandey, "Sean Parker: Facebook Was Designed to Exploit Human
'Vulnerability,'" *Axios*, November 9, 2017.

21. Amy B. Wang, "Former Facebook VP Says Social Media Is Destroying Society with 'Dopamine-Driven Feedback Loops,'" *Washington Post*, December 12, 2017.

22. Adrian F. Ward, interview with the author, July 25, 2019.

23. Jacques Barzun, *From Dawn to Decadence: 500 Years of Western Cultural
Life* (New York: HarperCollins, 2000), xiv–xv.

24. Betsy Sparrow, Jenny Liu, and Daniel M. Wegner, "Google Effects on Memory: Cognitive Consequences of Having Information at Our Fingertips,"

Science, 333, no. 6043 (August 5, 2011): 776–78. For more evidence of how Internet use encourages people to bypass their own memory, see Benjamin C. Storm, Sean M. Stone, and Aaron S. Benjamin, "Using the Internet to Access Information Inflates Future Use of the Internet to Access Other Information," *Memory*, 25, no. 6 (2017): 717–23.

25. See, for example, Keith E. Stanovich, "Miserliness in Human Cognition: The Interaction of Detection, Override and Mindware," *Thinking & Reasoning*, 24, no. 4 (2018): 423–44.

26. Linda A. Henkel, "Point-and-Shoot Memories: The Influence of Taking Photos on Memory for a Museum Tour," *Psychological Science*, 25, no. 2 (2014): 396–402.

27. Diana I. Tamir, Emma M. Templeton, Adrian F. Ward, and Jamil Zaki, "Media Usage Diminishes Memory for Experiences," *Journal of Experimental Social Psychology*, 76 (2018): 161–68.

28. Henry H. Wilmer, Lauren E. Sherman, and Jason M. Chein, "Smartphones and Cognition: A Review of Research Exploring the Links Between Mobile Technology Habits and Cognitive Functioning," *Frontiers in Psychology*, 8 (April 2017), https://doi.org/10.3389/fpsyg.2017.00605.

29. Daniel M. Wegner and Adrian F. Ward, "How Google Is Changing Your Brain," *Scientific American*, December 2013.

30. Matthew Fisher, Mariel K. Goddu, and Frank C. Kell, "Searching for Explanations: How the Internet Inflates Estimates of Internal Knowledge," *Journal of Experimental Psychology: General*, 144, no. 3 (2015): 674–87. See also Ward, "Supernormal."

31. Soroush Vosoughi, Deb Roy, and Sinan Aral, "The Spread of True and False News Online," *Science*, 359, no. 6380 (March 9, 2018): 1146–1151.

32. Cynthia Ozick, "T. S. Eliot at 101," *New Yorker*, November 20, 1989.

Further Reading

This book scratches many surfaces. To the reader who would like to explore the topics further, I recommend the following books, all of which I found illuminating and many of which I found inspiring.

THE BRAIN AND ITS PLASTICITY

Buller, David J. *Adapting Minds: Evolutionary Psychology and the Persistent Quest for Human Nature.* MIT Press, 2005.

Cowan, Nelson. *Working Memory Capacity.* Psychology Press, 2005.

Doidge, Norman. *The Brain That Changes Itself: Stories of Personal Triumph from the Frontiers of Brain Science.* Penguin, 2007.

Dupuy, Jean-Pierre. *On the Origins of Cognitive Science: The Mechanization of the Mind.* MIT Press, 2009.

Flynn, James R. *What Is Intelligence? Beyond the Flynn Effect.* Cambridge University Press, 2007.

Golumbia, David. *The Cultural Logic of Computation.* Harvard University Press, 2009.

James, William. *The Principles of Psychology.* Holt, 1890.

Kandel, Eric R. *In Search of Memory: The Emergence of a New Science of Mind.* Norton, 2006.

Klingberg, Torkel. *The Overflowing Brain: Information Overload and the Limits of Working Memory.* Oxford University Press, 2008.

LeDoux, Joseph. *Synaptic Self: How Our Brains Become Who We Are.* Penguin, 2002.

Martensen, Robert L. *The Brain Takes Shape: An Early History.* Oxford University Press, 2004.

Schwartz, Jeffrey M., and Sharon Begley. *The Mind and the Brain: Neuroplasticity and the Power of Mental Force.* Harper Perennial, 2002.

Sweller, John. *Instructional Design in Technical Areas.* Australian Council for Educational Research, 1999.

Wexler, Bruce E. *Brain and Culture: Neurobiology, Ideology, and Social Change.* MIT Press, 2006.

Young, J. Z. *Doubt and Certainty in Science: A Biologist's Reflections on the Brain.* Oxford University Press, 1951.

THE HISTORY OF THE BOOK

Chappell, Warren. *A Short History of the Printed Word.* Knopf, 1970.

Diringer, David. *The Hand-Produced Book.* Philosophical Library, 1953.

Eisenstein, Elizabeth L. *The Printing Press as an Agent of Change.* Cambridge University Press, 1980. An abridged edition, with a useful afterword, has been published as *The Printing Revolution in Early Modern Europe* (Cambridge University Press, 2005).

Kilgour, Frederick G. *The Evolution of the Book.* Oxford University Press, 1998.

Manguel, Alberto. *A History of Reading.* Viking, 1996.

Nunberg, Geoffrey, ed. *The Future of the Book.* University of California Press, 1996.

Saenger, Paul. *Space between Words: The Origins of Silent Reading.* Stanford University Press, 1997.

THE MIND OF THE READER

Birkerts, Sven. *The Gutenberg Elegies: The Fate of Reading in an Electronic Age.* Faber and Faber, 1994.

Dehaene, Stanislas. *Reading in the Brain: The Science and Evolution of a Human Invention.* Viking, 2009.

Goody, Jack. *The Interface between the Written and the Oral.* Cambridge University Press, 1987.

Havelock, Eric. *Preface to Plato.* Harvard University Press, 1963.

Moss, Ann. *Printed Commonplace-Books and the Structuring of Renaissance Thought.* Oxford University Press, 1996.

Olson, David R. *The World on Paper: The Conceptual and Cognitive Implications of Writing and Reading.* Cambridge University Press, 1994.

Ong, Walter J. *Orality and Literacy: The Technologizing of the Word.* Routledge, 2002.

Wolf, Maryanne. *Proust and the Squid: The Story and Science of the Reading Brain.* Harper, 2007.

MAPS, CLOCKS, AND SUCH

Aitken, Hugh G. J. *The Continuous Wave: Technology and American Radio, 1900–1932.* Princeton University Press, 1985.

Harley, J. B., and David Woodward, eds. *The History of Cartography,* vol. 1. University of Chicago Press, 1987.

Headrick, Daniel R. *When Information Came of Age: Technologies of Knowledge in the Age of Reason and Revolution, 1700–1850.* Oxford University Press, 2000.

Landes, David S. *Revolution in Time: Clocks and the Making of the Modern World,* rev. ed. Harvard University Press, 2000.

Robinson, Arthur H. *Early Thematic Mapping in the History of Cartography.* University of Chicago Press, 1982.

Thrower, Norman J. W. *Maps and Civilization: Cartography in Culture and Society.* University of Chicago Press, 2008.

Virga, Vincent, and the Library of Congress. *Cartographia: Mapping Civilizations.* Little, Brown, 2007.

TECHNOLOGY IN INTELLECTUAL HISTORY

Heidegger, Martin. *The Question concerning Technology and Other Essays.* Harper & Row, 1977. Heidegger's essay on technology was originally published in the collection *Vorträge und Aufsätze* in 1954.

Innis, Harold. *The Bias of Communication.* University of Toronto Press, 1951.

Kittler, Friedrich A. *Gramophone, Film, Typewriter.* Stanford University Press, 1999.

Marx, Leo. *The Machine in the Garden: Technology and the Pastoral Ideal in America.* Oxford University Press, 2000.

McLuhan, Marshall. *The Gutenberg Galaxy: The Making of Typographic Man.* University of Toronto Press, 1962.

McLuhan, Marshall. *Understanding Media: The Extensions of Man*, critical ed. Gingko, 2003.

Mumford, Lewis. *Technics and Civilization.* Harcourt Brace, 1934.

Postman, Neil. *Technopoly: The Surrender of Culture to Technology.* Vintage, 1993.

COMPUTERS, THE INTERNET, AND ARTIFICIAL INTELLIGENCE

Baron, Naomi S. *Always On: Language in an Online and Mobile World.* Oxford University Press, 2008.

Crystal, David. *Language and the Internet*, 2nd ed. Cambridge University Press, 2006.

Dyson, George B. *Darwin among the Machines: The Evolution of Global Intelligence.* Addison-Wesley, 1997.

Jackson, Maggie. *Distracted: The Erosion of Attention and the Coming Dark Age.* Prometheus, 2008.

Kemeny, John G. *Man and the Computer.* Scribner, 1972.

Levy, David M. *Scrolling Forward: Making Sense of Documents in the Digital Age.* Arcade, 2001.

Von Neumann, John. *The Computer and the Brain*, 2nd ed. Yale University Press, 2000.

Wiener, Norbert. *The Human Use of Human Beings.* Houghton Mifflin, 1950.

Weizenbaum, Joseph. *Computer Power and Human Reason: From Judgment to Calculation.* Freeman, 1976.

Acknowledgments

This book grew out of an essay I wrote for the *Atlantic* called "Is Google Making Us Stupid?," which appeared in the magazine's July–August 2008 issue. I thank the *Atlantic*'s James Bennet, Don Peck, James Gibney, Timothy Lavin, and Reihan Salam for their help and encouragement. My discussion of Google's strategy in chapter 8 draws on material that originally appeared in "The Google Enigma," an article I wrote for *Strategy & Business* in 2007. I am grateful to Art Kleiner and Amy Bernstein at that magazine for their expert editing. The afterword draws on my 2017 *Wall Street Journal* essay "How Smartphones Hijack Our Minds." For their generosity in taking time to answer my questions, I thank Mike Merzenich, Maryanne Wolf, Jim Olds, Russell Poldrack, Gary Small, Ziming Liu, Clay Shirky, Kevin Kelly, Bruce Friedman, Matt Cutts, Tom Lord, Caleb Crain, Bill Thompson, and Ari Schulman. I owe particular thanks to my editor at W. W. Norton, Brendan Curry, and his talented colleagues. I am also indebted to my agent, John Brockman, and his associates at Brockman Inc. Finally, I salute the book's intrepid first readers: my wife Ann and my son Henry. They made it to the end.

Index

abacus, 44
abbreviations, 252n
abstract thinking, 43–44, 50, 140, 147
academia, 8–9, 111–12, 138, 153–54,
    217–18
accounting tokens, 52
adaptation, 19–36, 211–12, 221–22
Adapting Minds (Buller), 31
addiction, 35, 116, 233–34
adrenaline, 35
advertising, 95, 117, 154–60, 163
Adweek, 88
AdWords, 155–57, 159
agrarian rhythms, 41
agriculture, 41, 44, 45, 48, 210
Aiken, Conrad, 209
alerts, message, 132–34
algorithms, 84, 149–50, 152, 172, 175,
    181, 202–5, 213, 216, 218–19,
    223–24
alienation, 166–67, 168, 171, 209–10,
    212, 219–20
All Citizens Are Soldiers (Vega), 71
alphabet, 51, 53–54, 55, 57, 67, 68, 116
Alzheimer's disease, 242n
Amazon.com, 15, 94, 101, 218
ambiguity, 151–52, 173
Ambrose, Saint, 60–62

American Association for the Advance-
    ment of Science, 171–72
American Library Association, 97, 163
America Online (AOL), 14
Amish, 47
amnesia, 183
amplification, physical, 209–10
amplifiers, 44, 78–80
amputations, 29–30, 212
Analogue Youth, 10–11
"analytical engines," 82
anatomy, 36–37
Anatomy of Melancholy, An (Burton),
    168
Anders, Günther, 173
Anderson, Sam, 140
Android operating system, 159
Animal Farm (Orwell), 252n
"animal spirits," 37
animation, 84
annotations, 178
Annual Review of Sociology, 108
anterior hippocampus, 33
antibiotics, 184
anti-intellectualism, 111–12
Aplysia (sea slug), 27–28, 182, 184–88
Apple Computer, 11, 12–14, 85, 92
Apple Macintosh computers, 12–13

Apple Macintosh Performa 550 computer, 14
Apple Macintosh Plus computer, 12–14
appliances, 79
"apps" (software applications), 91, 132, 157–58, 159, 195, 227, 233–34
Aristotle, 36–37
art, 75
artifacts, human, 102
artificial intelligence (AI), 5, 10, 16, 152–53, 172–76, 181–82, 191, 201–8, 223–24
artificial memory, 181–82
Association of American Publishers, 162
associative indexing, 169–70
astrocytes, 241n
"As We May Think" (Bush), 169–70
Atkinson, Bill, 13, 170
Atlantic Monthly, 109, 169, 222
attention deficit disorder (ADD), 125, 222, 229
attention restoration theory (ART), 219
attention span, 102, 125, 136–43, 165–66, 193–97, 219–20, 230–31, 233–34, 246n
Au, Irene, 151, 156
audio, streaming, 78–80, 91–92, 116–17, 129–34
Audion, 78–80
audiotapes, 180
auditory cortex, 116–17, 189
auditory sense, 29, 53, 110, 116–17, 131, 189
Augustine, Saint, 61–62, 181
Austen, Jane, 76
Australia, 136
Authors Guild, 162
autoamputation, 212
automated information filtering, 216–19
automated online tests, 151
automation, 149, 207, 214–19
automatons, 149
automobiles, 47
Axel, Richard, 186–87
axons, 19–20, 27, 194

Babbage, Charles, 82
Baby Boomers, 10–11
backlit screens, 100–101
BackRub, 154
Bacon, Francis, 69, 70, 72, 179
Baker Library, 11, 12
bandwidth, 83, 97, 98, 146, 154, 198, 200
Baron, Naomi, 179–80
Basel, University of, 17

BASIC, 11, 12
Battelle, John, 10
Beatles, 1, 10
Beethoven, Ludwig van, 82, 97
behavior, 21, 34–35, 150–76, 187, 215
Bell, Daniel, 44
Bell, David, 103–4
Bell, Vaughan, 64
Bellamy, Edward, 109
bells, 42
Benedict, Saint, 42
Berman, Marc, 219–20
Berners-Lee, Tim, 9–10
Bernstein, Michael, 30
Bezos, Jeff, 108
Bible, 60, 68, 94, 210–11
bibliophiles, 110
binary bits, 94, 190–91, 199
Bing search engine, 216
biological memory, 181–93
birth control pill, 44
BlackBerrys, 15
blindness, 29, 30, 49 65–66
blogs, 6, 7–8, 9, 15, 85, 97, 111, 117, 132, 151, 157, 159, 165, 198
blue, 151
Blu-ray, 96, 200
Bodkin, Tom, 95
Bodleian Library, 162
bold type, 117
bookmarks, 101
books, 5–8, 11, 15–16, 44, 50–77, 73, 87–95, 97, 98–114, 116, 120, 122–29, 131, 134–38, 140, 141, 157, 161–66, 168, 170–71, 177–78, 199, 246n, 247n, 252n, 262n
    see also publishing industry
"Books" (Emerson), 170–71
booksellers, 69–71
Bos, Maarten, 229
Boswell, James, 142–43
bottom-up mechanisms, 63–64
"bouncing out," 136
Bowen, Elizabeth, 127
boxing, 183
Boyd, Danah, 86
Braille, 29, 49
brain:
    Aristotle's theory of, 36–37
    "artificial" (artificial intelligence), 5, 10, 16, 152–53, 172–76, 181–82, 191, 201–8, 223–24

autonomy of, 26–28, 175–76
behavior and, 21, 34–35, 187
cellular change in, 49, 182–88,
    241n–42n
cerebral cortex of, 24–26, 29, 31–32,
    53, 61, 63, 116–17, 122, 153, 188,
    189–90, 213, 246n, 262n–63n
computers compared with, 23, 26,
    34, 152–53, 172–76, 180–82,
    190–93, 199, 201–8
development of, 16, 20–21, 31, 32,
    38, 39–43, 48–49, 115–16, 119–20,
    139–48, 182–93, 208–12, 221–22
diseases of, 23–24, 183, 188–89, 242n
evolution of, 31, 48–49
frontal lobe of, 51–52, 194
genetic influences on, 28–29, 31, 34,
    35, 49, 171–72, 187, 190
gray matter of, 49
intellectual development and, 34–35,
    39–45, 64–65, 71–72, 75–76,
    111–12, 144–48, 184–88, 191–97,
    219–20
language centers of, 50–77
learning by, 184–88, 191–92
maps of, 24–26, 32
mechanistic conception of, 22–26,
    28–29, 34, 37, 172, 173–76
memory capacity of, 20–21, 32–33,
    44, 49–57, 121–26, 177–97, 199,
    219, 242n, 262n–63n
mind compared with, 22–23, 32–35,
    37, 48–50, 175–76, 195–97
neurons of, 5, 19–35, 48–52, 53, 63,
    71–72, 77, 119–23, 139–40, 141,
    175–76, 182–94, 199, 211–12, 213,
    221, 240n–42n, 246n
organic model for, 179, 181–87,
    190–93, 195
plasticity of, 21–35, 48–50, 75, 115,
    119–20, 139–43, 176, 182–88,
    192–97, 211–12, 213, 221–22
regeneration of, 49–50
research on, 26, 34–35, 37, 51–52,
    63–64, 74, 120–22, 126–29,
    134–35, 139–43, 175, 182–97, 213,
    214–16, 220–21, 240n–41n
scans of, 26, 37, 51–52, 74, 175, 213,
    220–21
spatial representation by, 32–33,
    39–41, 49–50, 53, 211–12,
    262n–63n
structure of, 20–21, 24–35, 37,
    48–50
synapses in, 19–22, 26–27, 34–35,
    49, 71–72, 120, 141, 182–93, 194,
    199, 240n–42n
temporal lobe of, 51–52, 213
temporal representation by, 41–44,
    45, 50, 70, 211
trauma to, 30–31, 34, 183, 188–89
    see also specific parts
Brain and Creativity Institute, 220
"Brain Drain," 229, 230
Brain Takes Shape, The (Martensen), 37
Brain That Changes Itself, The (Doidge),
    30, 34, 35
Brin, Sergey, 154–56, 161–62, 172–73, 174
British Library, 136
broadband, 15, 83, 97, 98, 146, 154, 198,
    200
broadcasts, television, 95–96, 112
Bronx Library Center, 97–98
Brooks, David, 180–81, 182
Brooks, Tim, 93
browsers, Internet, 14–15, 85, 121–22,
    130, 135–38, 195
Buller, Dave, 31
bulletin boards, electronic, 14
bureaucracies, 168–69
Bureau of Labor Statistics, U.S., 87–88
Burroughs, William, 76
Burton, Robert, 168
Bush, Vannevar, 169–70
business, 58, 94, 95–96, 117, 154–60, 163

cable television, 96, 131
Caffeine search architecture, 159
calculators, 83, 88, 180, 192–93
calendars, 159, 215
Cambridge, Richard Owen, 142–43
Cambridge University, 81, 162
Cambridge University Press, 162
cameras, 89
Canada, 136
capitalism, 46, 89
carbon atoms, 241n
Carey, James, 46
carrels, 66
Carroll, James, 67
Carson, Rachel, 76
Cartesianism, 22–23, 33, 37, 50, 70, 76
cartography, 39–41, 44, 45, 47, 49–50,
    83, 206, 211–12

cast letters, 68
catalytic enzymes, 186
Catholic Church, 68
CD burners, 14
CD-ROM drives, 14
CDs, 14, 89, 92
"cell phone novels," 104–5
cell phones, 15, 86, 92, 94, 96–97,
    104–5, 116, 146, 159
cells, brain, 49, 182–88, 241n–42n
    see also neurons
cenobites, 62
censorship, 71
Center for Media Design, 87
central nervous system, 176, 213
cerebral cortex, 24–26, 29, 31–32, 53,
    61, 63, 116–17, 122, 153, 188,
    189–90, 213, 246n, 262n–63n
Cervantes Saavedra, Miguel de, 70
Chabris, Christopher, 134
chamber clocks, 43
character recognition programs, 161
chatter, 106–7
Chicago Tribune, 93
China, 69, 86
Chinese language, 51, 89, 246n
chips, computer, 23, 83, 180
Chomsky, Noam, 202
Christianity, 42, 60–68
Christian Science Monitor, 94
Cicero, Marcus Tullius, 70
cinema, 5, 24, 77, 84, 85, 88, 172–73,
    175, 224
circulation, magazine, 95
Cistercian monks, 42
citations, scholarly, 153–54, 217–18
classical literature, 55, 70, 178–79
classification, 147
classroom lectures, 66
clay tablets, 52–53, 58–59
clepsydras, 41
ClickTale, 136
clocks, 41–44, 47, 50, 70, 83, 92, 172,
    206, 211, 212, 244n
cloisters, 66
closure, 106–7
cloth manufacturing, 210
cloud computing, 83
CNN, 131
code, software, 83
codex, 60–61, 67, 109–10
cognitive development, 5–10, 39–45,

49–50, 64–65, 71–76, 90, 104,
    105–6, 111–12, 119–33, 136–43,
    147–52, 165, 169–70, 173,
    192–200, 214–21
cognitive load, 125–29, 133
cognitive tests, 144–48, 219–20
collaborative online filters, 180
collective memory, 196
colored-ball puzzle, 214–16
Columbia University, 186
commercial economies, 147
commonplace books, 179–80
communal societies, 196, 213
compact discs (CDs), 14, 89, 92
"complements," 159–60
complexity, 123–26, 151–52, 196
complex memories, 188
comprehension, 126–34
CompuServe, 14
computational linguists, 202
Computer Power and Human Reason
    (Weizenbaum), 206–7
computers:
    author's experience with, 5–7, 11–16,
        38
    brain compared with, 23, 26, 34,
        152–53, 172–76, 180–82, 190–93,
        199, 201–8
    cost of, 83, 92
    cultural influence of, 5–7, 11–16, 38,
        77, 101–7, 201–19
    desktop, 77
    development of, 11–12, 14, 81–83,
        113, 180, 208–19
    digital, 79, 81–82, 83, 88–89, 93,
        207, 214, 248n, 252n–53n
    in education, 92–93, 126–29,
        214–16, 223
    handheld or mobile, 77, 96–97, 101,
        102–5, 116, 117
    hard drives of, 12, 14, 15, 180, 182
    hardware for, 12–13, 14, 83, 170, 216
    human interaction with, 11–12, 16,
        180–81, 191–97, 201–19
    as intellectual technology, 23, 44–45,
        180–81, 191–97, 201–19, 221–22
    interface for, 113, 216–17
    keyboards for, 12, 116, 169, 209–10
    laptop, 77, 92
    mainframe, 11, 79, 83, 201
    memory capacity of, 83, 101, 181–82,
        190–91, 194

mice (pointing devices) for, 12, 116
microprocessors for, 14, 83, 216
modems for, 13–14, 15
multitasking on, 113, 132–33,
     139–43, 221
networks of, 11–14, 82–83, 170, 213;
     see also Internet
office, 132
operating systems for, 14, 113, 159,
     172
personal (PCs), 4, 11–16, 83, 88, 113,
     144, 148, 169–70, 172, 260n
programming language for, 11, 12, 201
prototypes of, 81–82, 102, 169–70
public access to, 11–12, 97–98,
     202–5
screens (monitors) for, 12, 14, 77, 87,
     94, 100, 103, 107, 116, 138, 169
software for, 13–14, 82, 83, 84,
     91, 100, 107, 113, 132, 135, 150,
     157–58, 159, 170, 175, 195, 201–8,
     214–19
terminals for, 11–12, 201
text displayed by, 99–114
time-sharing for, 11–12, 204–5, 207
universality of, 81–83, 102, 213
computer science, 152–53
"Computing Machinery and Intelli-
     gence" (Turing), 82, 174, 205–6
concentration, mental, 5–8, 64–65,
     90–91, 94–98, 102–4, 110–43,
     156–57, 165–71, 193–200, 219–22,
     246n
conceptualization, 44–45, 50, 52,
     121–26, 140, 147
Concord, Mass., 166–67, 171
concordances, 66, 90, 168
Confessions (St. Augustine), 60
consciousness, 22–23, 50–51, 56–57, 67,
     76, 119, 120, 123–24, 166–67, 168,
     171, 193–94, 219–20, 265n
consolidated memories, 183–84,
     190–93, 194, 262n–63n
consumer preference algorithms, 181
"contact barriers," 19, 20, 21–22
contemplation, 65, 120, 166–67, 168,
     171, 219–20, 222, 265n
contextual ads, 155
Convent of San Jacopo di Ripoli, 69
copyright, 162–66
Copyright Office, U.S., 163
Cornell University, 130

corn plants, 44
correlation, 147
correspondence, personal, 84, 92, 107–8
     see also e-mail
Cowan, Nelson, 192
Cowen, Tyler, 94
Crain, Caleb, 107
crawls, text, 95–96
creativity, 119, 140, 192–93, 218,
     220–21
CREB-1, 187
CREB-2, 187
critical reading scores, 145–46
crossword puzzles, 122, 125–26
Crovitz, L. Gordon, 102
Crowell, Sheila, 192
Culkin, John, 210, 214
"cultural capital," 108
Cultural Logic of Computation, The
     (Golumbia), 202
cuneiform, 52–53, 58–59
Curr, Judith, 105–6
Current Biology, 63–64
cursive script, 209–10
Cushing Academy, 260n
cut-and-paste tools, 165
cyclic AMP, 186
cytoplasm, 187

Damasio, Antonio, 220–21
Darnton, Robert, 72, 163–64, 165, 247n
Dartmouth College, 11–12, 174–75
Dartmouth Time-Sharing System,
     11–12
Darwin, Charles, 76
Darwin among the Machines (Dyson),
     173–74
databases, 16, 82–83, 133–34, 153–54,
     169, 180, 192–93
data mining, 154, 165–66
data-processing technology, 16, 82–83,
     133–34, 153–54, 166–70, 180,
     192–93, 208
Davis, Philip, 7–8
"Dawn of the Electronic Age" (de For-
     est), 79–80
deafness, 29, 30
deciphering text, 63, 64, 81, 166
decision making, 53, 121–22, 129, 151, 223
Decline and Fall of the Roman Empire,
     The (Gibbon), 76
de Cou, Emil, 96

"deep processing," 141

deep reading, 5–7, 60–67, 73–74, 97, 106–11, 122–29, 136–38, 140, 141, 166, 246n

de Forest, Lee, 78–80

Delany, Paul, 126

delusional thinking, 205

"Demon Lover, The" (Bowen), 127

dendrites, 19–20, 27

Denmark, 146

depression, 35

Descartes, René, 22–23, 33, 37, 50, 70, 76

design, book, 102

Desimone, Robert, 246n

determinism, 31, 34, 46–48

Developing Mind, The (Siegel), 190

Dialogues (Plato), 54–57, 69

Dickens, Charles, 103

dictation, 65–66, 265n

diction, 56, 75

dictionaries, 66

Digital Adulthood, 10–11

"digital immersion," 9

digital publications, 8–9, 94, 98–114, 126–38, 161–66, 252n, 260n

digital technology, 8–9, 79, 81–82, 83, 88–89, 94, 100, 101, 160, 161–66, 190–91, 199, 207, 214, 248n, 252n–53n

Dijksterhuis, Ap, 119

diodes, 78

diseases, brain, 23–24, 183, 188–89, 242n

Disney, 96

Distracted (Jackson), 133

distraction, 64, 102, 111–29, 133, 136–43, 156–57, 165–66, 193–200, 219–22, 246n

"divided attention," 125–26

DNA, 171–72

Doctorow, Cory, 91

Doidge, Norman, 23–24, 209, 213, 214

domain names, 15

dopamine, 35, 194, 241n–42n

dot-coms, 154, 157

drawing, 39–40

Droid, 92

drug addiction, 35

DSL, 198

dualism, 22–23, 33, 37, 50, 70, 76

Dumont, Léon, 21, 35

Duke, Kristen, 229

DVDs, 15, 89, 92

Dynamic Cognition Laboratory, 74

dynamic pages, 157

Dyson, George, 173–74

eBay, 15

Ebbinghaus, Hermann, 182–83, 184

e-books, 8–9, 73, 94, 98–114, 126–38, 161–66, 252n, 260n

Eco, Umberto, 178

Edexcel, 223

"edges," 104

Edison, Thomas, 109, 110

editions, published, 69–70

education, 8–9, 44, 81, 92–93, 126–31, 136–37, 145, 146, 147, 148, 162, 163, 168–69, 180, 192–93, 214–16, 223

educational testing, 145, 146, 223

Education Department, U.S., 146

efficiency, 137, 149–50, 152, 165–66, 168, 172–73, 216–17, 218

Egyptian hieroglyphics, 52–53, 59

E Ink, 100

Einstein, Alfred, 76

Eisenstein, Elizabeth, 69–72, 75

electricity, 78–80, 110

electronic distribution, 92, 128–38

"electron physiologists," 80

electrons, 79

Eliot, T. S., 64, 119, 209

ELIZA program, 202–5, 213

e-mail, 6, 9, 14, 16, 84, 91, 107–8, 113, 117, 132, 159, 198–99, 200

Emerson, Ralph Waldo, 46, 48, 76, 166, 170–71

emoticons, 252n

empathy, 213, 220–21

empiricism, 28–29

encyclopedias, 88

"End of Books, The" (Uzanne), 109–10

Engelbart, Douglas, 170

English language, 52, 56, 61, 62, 63, 75, 107, 201, 253n

Enigma machine, 81

Enlightenment, 10, 23, 37, 43, 181

entertainment industry, 92

entrepreneurship, 154–55, 171–72

enzymes, 186, 187

epilepsy, 183, 188–89

Erasmus Desiderius, 178–79

essays, 95

estrogen, 241n–42n
Euphrates River, 52
Europe, 43–44, 86
    see also specific countries
Evans, James, 217
Everything Bad Is Good for You (Johnson), 122–23
evolution, brain, 31, 48–49
evolution, theory of, 31, 76
evolutionary psychology, 31
eWorld, 14
exams, 223
Excel program, 13
explicit memories, 187–89, 194
external drives, 14
"extraneous problem-solving," 125–26
eye fatigue, 100–101
eye fixations, 134
eye tracking, 151

Facebook, 6, 16, 85, 91, 96, 98, 117, 118, 158, 198, 226, 233–34
factories, 149–50, 172, 218
fair-use provisions, 164
Federman, Mark, 111–12
feed readers, 132
fiber-optic cables, 84
fiction, 74, 103, 104, 127–28
Fight for Glorton game, 98
film cameras, 89
flat-screen televisions, 260n
Flaubert, Gustave, 76
Flexner, Louis, 184
Flickr, 85
flip books, 84
"flippers," 95–96
Florence, 69
Flynn, James, 144, 146–47, 148, 257n
Flynn effect, 144–45
folio, 70
Foreman, Richard, 196, 220
forgetting, 54–55, 182–84, 188–89, 193–97
Four Quartets (Eliot), 64, 119
fragmentation, mental, 90–91, 94–98, 129–34
France, 69, 70, 134, 136
Frankfurt Book Fair, 161–62
freelance writers, 15
"freshness," 158–59
Freud, Sigmund, 19, 20, 21–22, 26, 27
Frey, Scott, 209
Friedman, Bruce, 7, 8

From Down to Decadence (Barzun), 234
frontal cortex, 153
frontal lobe, 51–52, 194
Fust, Johann, 68–69, 70

games, computer, 12, 13, 85, 98, 139, 144
gamma-aminobutyric acid (GAMA), 241n–42n
Gargantua (Rabelais), 70
Geiger counters, 44
General Electric GE-635 mainframe computer, 11
General Theory of Employment, Interest and Money (Keynes), 76
Generation Net, 9
Generation X, 10–11
Genesis of Napoleonic Propaganda, The, 103–4
genetics, 28–29, 31, 34, 35, 49, 171–72, 187, 190
Genoa, Italy, 17–19
genome, 31
Germany, 68–69, 135–36
Gibbon, Edward, 76
glial cells, 240n–41n
global positioning system (GPS), 212
globes, 44
glossaries, 66
glutamate, 187, 241n
Gneezy, Ayelet, 229
God, 42, 50, 174, 181, 251n
Golumbia, David, 202
Goody, Jack, 44, 247n
Google, 6, 8–9, 15, 16, 91, 92, 94, 97, 101, 106, 117–18, 121, 138, 149–76, 181, 197, 216, 218, 234
Google Book Search, 8–9, 94, 161–66
Google Chrome, 159
"Google effect," 236
Googleplex, 150, 161, 174, 176
Google Print program, 161–62
Google Wave, 159
googol, 154
Gore-Tex, 212
Gothic typeface, 68
GoTo, 155
GPS, 212
Grafman, Jordan, 140
grand mal seizures, 188–89
graphical user interface (GUI), 113, 216–17
graphics, 84, 95
gray matter, 49

Great Britain, 62, 71, 136–37, 139, 146, 223
"Great Clockmaker," 50
Greek language, 53–57, 59, 71, 181–82
Greenberg, Michael, 34
Greenfield, Patricia, 141, 147
Greengard, Paul, 186
"groupiness," 107
Grown Up Digital (Tapscott), 144–45
Guardian, 93
Guardian News and Media, 93
Guibert of Nogent, 65–66
guitar amps, 79
Gutenberg, Johannes, 2, 10, 68–70, 72, 76, 83, 94, 110, 162, 168, 246n

habits, 21, 34–35
HAL 9000 computer, 5, 10, 16, 172–73, 175, 224
Hallett, Mark, 31
Hamlet (Shakespeare), 178
hand-eye coordination, 139
handheld devices, 77, 96–97, 101, 102–5, 116, 117
handwriting, 209–10
hard drives, 12, 14, 15, 180, 182
hardware, computer, 12–13, 14, 83, 170, 216
HarperCollins, 102
HarperStudio, 102–3
harrows, 210
Harvard University, 162
Hausauer, Michael, 118
Havelock, Eric, 55
Hawthorne, Nathaniel, 166–67, 168, 171, 219, 220
Hayles, Katherine, 9
headlines, 91, 95, 139, 157–58
heart, 36, 37
Hebb's rule, 27
Heidegger, Martin, 222
Henkel, Linda, 236–37
hieroglyphics, 52–53, 59
highlighted text, 101
"high reading," 110–11
high-speed connections, 146, 198, 200
Hillis, Danny, 244n
hippocampus, 32–33, 49–50, 188–90, 194, 211–12, 262n–63n
Hippocrates, 36–37
Homer, 55
Horace, 178

horizontal browsing, 136–37
Houghton Mifflin Co., 162
hourglasses, 41
"House Was Quiet and the World Was Calm, The" (Stevens), 73–74
Howard Hughes Medical Institute, 186
Hubert, Philip, 109
Huffington Post, 85
Hulu, 94
hunting, 48
HyperCard, 13, 170
hypermedia, 129–34, 170
hypertext links (hyperlinks), 13, 14–15, 90–91, 101–2, 117–18, 122, 126–38, 139, 153, 170, 214–18
hypertext literacy, 126–27

ideas, 44–45, 126–29, 177–78, 192–94
imagination, 75–76
"imitation game," 205–6
Immordino-Yang, Mary Helen, 221
implicit memories, 187–89
"In Defense of Distraction" (Anderson), 140
indexes, 90, 169–70, 181
India, 136
Indianapolis Symphony Orchestra, 96–97
"indispensables," 109
individuality, 67, 113–14, 138, 207–8, 212, 213–14, 218–22
indulgences, 68
Industrial Revolution, 10, 22, 149–50, 167–68, 209–12
information:
    bits of, 94, 190–91, 199
    comprehension of, 130, 214–19
    flow of, 83, 86, 93, 157–61, 221–22
    management of, 167–71, 192, 221–22
    monopoly on, 160–66
    overload of, 12, 125–29, 133, 170–71, 196–97
    storage of, 159, 180–82, 191–92
    technology for, 1–4, 44–45, 81–82, 116–20, 123, 138–43, 146
    transmission of, 1–4, 6–9, 85, 94, 130–43, 151–52, 157–59, 194, 196–97, 198, 221–22, 247n
Information Age, 79–80, 180–81, 221–22
Information and Control (Chomsky), 202
ink, 68, 100
innate talents, 51

In Search of Lost Time (Proust), 111
In Search of Memory (Kandel), 185
instant messages, 117, 118, 198
Institutional Subscription Database, 163
instrumentalism, 46–48
intellectual development, 34–35, 39–45, 64–65, 71–72, 75–76, 111–12, 144–48, 184–88, 191–97, 219–20
intellectual property, 162–66
intellectuals, 111–12
intelligence, 141, 143, 144–48, 172–73
    see also artificial intelligence
Internet:
    advertising on, 95, 117, 154–57, 159–60, 163
    audio on, 78–80, 91–92, 116–17, 129–34
    author's experience of, 5–7, 11–16, 38, 198–200, 223–24
    bandwidth (broadband) for, 15, 83, 97, 98, 146, 154, 198, 200
    behavioral data on, 150–76
    as bidirectional, 85–86
    blogs on, 6, 7–8, 9, 15, 85, 97, 111, 117, 132, 151, 157, 159, 165, 198
    brain development and use of, 16, 32, 38, 115–16, 119–20, 139–43
    browsers for, 14–15, 85, 121–22, 130, 135–38, 195
    concentration interrupted by, 5–8, 64–65, 90–91, 94–98, 102–4, 110–14, 121–26, 129–43, 165–71, 193–200, 219–22, 246n
    cultural influence of, 2–4, 9–10, 77, 85–98, 138, 194–97
    databases of, 16, 82–83, 133–34, 153–54, 169, 180, 192–93
    dependence on, 2–3, 5–7, 11–16, 94–98, 116–23, 132–34, 141–43, 198–200, 221–24
    development of, 9–10, 14–15, 83–85, 88–90, 153, 156
    economics of, 88–89, 140, 154–57, 159
    education and, 8–9, 129–31, 147, 148
    as electronic network, 8, 11–14, 82–83, 85, 170, 181, 213
    e-mail as service of, 6, 9, 14, 16, 84, 91, 107–8, 113, 117, 132, 159, 198–99, 200
    generational differences in, 8–11
    graphics on, 84, 95
    for handheld or mobile devices, 77, 96–97, 101, 102–5, 116, 117
    high-speed connection to, 146, 198, 200
    hypertext links (hyperlinks) on, 13, 14–15, 90–91, 101–2, 117–18, 122, 126–38, 139, 153, 170, 214–18
    information content of, 1–4, 6–9, 85, 94, 130–43, 151–52, 157–59, 194, 196–97, 198, 221–22, 247n
    in intellectual technology, 44–45, 138–48, 150–52, 161–66, 199–200
    as medium, 1–4, 6, 9–10, 85–86
    as multimedia source, 14, 88–93, 131, 214–16
    music on, 84, 85, 88, 94
    as "network of networks," 14
    news sites on, 9, 157–58
    online services for, 14
    overstimulation by, 16–23, 132–34, 141–43
    profits from, 95, 154–60, 163
    psychological research on, 150–76
    public access to, 14, 97–98
    research on, 85–88, 126–76, 214–16
    search engines for, 6, 90–91, 106, 132–33, 181, 216; *see also* Google
    shopping on, 9, 89, 130
    as social network, 6, 85–86, 91, 96, 97, 98, 106–7, 117–18, 132–34, 158, 159, 198, 200, 248n, 251n
    software applications ("apps") on, 91, 132, 157–58, 159, 195
    speed of, 83, 86, 157–58
    statistical analysis of, 150–76
    storage capacity of, 159, 180–82, 191–92
    surfing of, 87, 96, 115–16, 121–22, 130–31, 150–76
    television compared with, 86–87, 88, 94, 95–96
    text displayed by, 77, 90, 91–92, 105–6, 128–38, 139, 161–66, 217–18
    as universal medium, 88–92
    video on, 84–85, 91–92, 116–17, 129–34, 200
    Web design on, 115–16
    Web pages on, 83–84, 91, 115–16, 120, 135–36, 157, 159, 218–19

Internet (*continued*)
  Web sites on, 9, 83–84, 91, 93, 94,
    120, 135–36, 153, 154, 157–58, 159,
    165, 216–19
  as World Wide Web, 9–10, 13, 14–15,
    83, 112, 127, 153, 170
  see also computers
interneurons, 186–87
iPads, 104, 116
iPhones, 4, 92, 102, 116, 225–26, 228
iPods, 15, 101, 116
IQ scores, 144–48
Iraq, 52
Ireland, 62
Isaac of Syria, 65
Isidore of Seville, Saint, 178
ISP accounts, 15
Italian language, 52
Italy, 52, 69
iTunes, 94

Jackson, Maggie, 133
James, Henry, 76
James, William, 21, 26, 181, 183, 195–96
Japan, 47, 104–5
Javal, Louis Émile, 134
Johnson, Samuel, 142–43
Johnson, Steven, 106, 122–23
journalism, 93
Journal of Nervous and Mental Disease,
    205, 206
journals, 70, 92, 153–54, 168, 177, 217–18
Joyce, James, 76
junk ads, 155
Jupiter Research, 87
Justice Department, U.S., 163

Kandel, Eric, 27–28, 182, 184–88, 191,
    193
Kansas State University, 131
Kant, Immanuel, 28, 76
Kanwisher, Nancy, 29
Karp, Scott, 7, 8
Keats, John, ix
Kelly, Kevin, 106, 108
Kemeny, John, 11–12
keyboards, 12, 116, 169, 209–10
Keynes, John Maynard, 76
keypads, 101, 209–10
keywords, 117, 138
Kiewit Computation Center, 11–12
kinase A, 186, 187

Kindle reader, 101–4, 108, 252n, 260n
kineographs, 84
Klingberg, Torkel, 118, 125, 192
knowledge, 55–57, 67, 128–29, 138,
    142–43, 145, 150, 163, 184–88,
    191–92
Koman, Richard, 164
Kornblau, Craig, 96
Köselitz, Heinrich, 18–19
Kubrick, Stanley, 5, 172–73, 175
Kuhn, Thomas, 76
Kunkel, Benjamin, 199
Kurzweil, Ray, 175

Lamartine, Alphonse de, 109
Landes, David, 42, 43
Landow, George, 126
"landscape" mode, 116
language, 50–77, 121–22, 145–46, 176,
    208, 223, 252n–53n
laptop computers, 77, 92
Late Night with Jimmy Fallon, 96
Latin language, 53, 59–61, 62, 71
lawsuits, copyright, 162–66
Leaves of Grass (Whitman), 196
lectio divina (holy reading), 246n
LeDoux, Joseph, 28–29, 191
Le Goff, Jacques, 41
letterpress, 68–70, 76, 177
letters, 84, 92, 107–8
Levy, David, 73, 113–14, 170
libraries, 44, 66, 70, 83, 97–98,
    142–43, 161–69, 177–78, 181,
    217–18, 260n
linearity, 5–10, 104, 105–6, 111–12,
    126–29, 136–38, 165
linguistics, 202
Linux, 172
literacy, 50–77, 110–12, 126–27, 145–46,
    177–78, 257n
  see also reading
literary reading aptitude, 146
Liu, Ziming, 137–38
Locke, John, 28, 76
logic, 82, 169, 176, 214–19
logic puzzles, 214–16
logographic symbols, 51–52, 246n
London cab drivers, 32–33, 49–50, 188,
    212
longhand, 209–10
long-term memory, 123–26, 182–97
looms, 212

Lope de Vega, 71
Los Angeles Times, 93
Lynch, David, 11

McCandliss, Bruce, 193
McGrath, Charles, 102
McGraw-Hill Co., 162
Machine in the Garden, The (Marx), 167
McLuhan, Marshall, 1–2, 6, 10, 24, 46,
    56–57, 89, 102, 210, 212
magazines, 14, 16, 84, 87, 89, 90, 91–92,
    94–95, 101, 109, 138, 157
"Magical Number Seven, Plus or Minus
    Two, The" (Miller), 124
magnetic resonance imaging (MRI),
    220–21
magnetism, 48
Maguire, Eleanor, 212
Maho no i-rando, 104
mail, 83, 84, 92, 107–8
    see also e-mail
Mailer, Norman, 24
mainframe computers, 11, 79, 83, 201
Mali, 130
Malling-Hansen, Hans Rasmus Johann,
    17–18
Malling-Hansen Writing Ball, 17–19,
    45, 265*n*
Man and the Computer (Kemeny), 11–12
Mangen, Anne, 90
Manguel, Alberto, 112
manual dexterity, 210
manual labor, 149–50, 218
manufacturing, 149–50
Manutius, Aldus, 70
MAP enzyme, 187
mapping, brain, 24–26, 32
maps, 39–41, 44, 45, 47, 49–50, 83, 206,
    211–12
Martensen, Robert, 37
Marx, Karl, 46
Marx, Leo, 167
Massachusetts Institute of Technology
    (MIT), 201–8
mass media, 1–2, 79–80, 83, 158
    see also radio; television
mathematics, 43–44, 82, 145, 146, 154,
    176, 192–93
Mathes, Adam, 165
math scores, 145, 146
Mayer, Marissa, 150, 151, 161
mechanical clocks, 41–44

mechanistic models, 22–26, 28–29, 34,
    37, 172, 173–76
media:
    bidirectional, 85–86
    corporations for, 94, 95–96
    digital, 79, 81–82, 83, 88–89, 93,
        100, 101–2, 106, 120–22, 137–38,
        146, 207, 214, 248*n*, 252*n*–53*n*
    electronic, 1–2, 77, 78–80, 89–90,
        92, 128–38, 140, 213
    forms of, 1–4, 6, 9–10, 46, 85–86, 89,
        210
    hyper-, 129–34, 170
    interactive, 96–97, 135–36
    mass, 1–2, 79–80, 83, 158
    multi-, 14, 88–93, 131, 214–16
    networks of, 8, 11–14, 82–83, 85,
        95–96, 170, 181, 213
    personal, 87–88
    "rich," 129–30
    social, 106–7
    traditional, 89–90
    see also specific media
medial temporal lobes, 188, 189
Media Psychology, 130
medical information, 139
meditation, 65, 120, 166–67, 168, 171,
    219–20, 222, 265*n*
Meditations (Descartes), 22–23
memex, 169–70
memorization, 32–33, 110–11, 145,
    178–79
memory, computer, 83, 101, 181–82,
    190–91, 194
memory, human, 20–21, 32–33, 44,
    49–57, 121–26, 177–97, 199, 219,
    242*n*, 262*n*–63*n*
Menon, Vinod, 232
mental illness, 23–24, 34–35
mental retardation, 146, 257*n*
Merzenich, Michael, 24–26, 30, 119–20,
    137, 142, 185
metallurgy, 48
metal type, 68
metalworking machines, 149–50
Mexico City, 70
Meyer, David, 140–41
mice (pointing devices), 12, 116
mice (rodents), 184, 193–94, 242*n*
Michigan, University of, 152–53,
    219–20
microblogs, 15, 97, 132

microelectrodes, 24–26
microfiche, 180
microfilm, 180
micromaps, 24–26
microprocessors, 14, 83, 216
microscopes, 19, 44, 45
Microsoft Corp., 13, 16, 91, 105, 216
Microsoft Word program, 13, 16
Middle Ages, 42–43, 60–62, 65–66, 69, 77, 181
Midvale Steel, 149
Miller, George, 124
Milner, Brenda, 189
Milton, John, 70
mind, 22–23, 32–35, 37, 48–50, 175–76, 195–97
    see also cognitive development
miniaturization, 43, 70, 92
Minneapolis Star, 93
mirroring, neural, 213
Mitchell, Jason, 213
Mnemosyne, 181–82
mobile devices, 15, 77, 86, 87, 91, 92, 94, 96–97, 101, 102–5, 116, 117, 146, 159
modems, 13–14, 15
Modernism, 10
modular architecture, 105
Molaison, Henry, 188–89, 262n
molecules, 186–87
Molière, 70
monitors, computer, 12, 14, 77, 87, 94, 100, 103, 107, 116, 138, 169
monkeys, 24–26, 30, 32, 208
monks, 42, 60–62, 65–66, 70
monographs, 92
monorails, 153
Montana, Hannah, 98
Motorola, Inc., 92
Mountcastle, Vernon, 24
movable type, 68–70, 89, 97, 246n
movies, 77
MP3 files, 15, 84, 101
"Mugby Junction" (Dickens), 103
Müller, Georg, 183–84
multimedia programs, 14, 88–93, 131, 214–16
multitasking, 113, 132–33, 139–43, 221
Mumford, Lewis, 44, 175, 211
Muses, 181
music, 75, 84, 85, 88, 89, 92, 94, 96–97

MySpace, 16, 85, 158

Napster, 15
Nass, Clifford, 142
National Institute of Neurological Disorders and Stroke, 140
National Symphony Orchestra, 96
Natural History, 205
natural language, 202, 204
natural selection, 31
Nature, 139
nature vs. nurture debate, 28–29, 31
Nazis, 81
NBC, 96
netbooks, 15, 92
Netflix, 85, 200, 218
Net Generation, 9
Netscape browser, 15
"networked thinking," 8, 181
networks, television, 95–96
Neurobiology of Learning, The, 192
neuroimaging, 213
"neurological nihilism," 23–24
neurons, 5, 19–35, 48–52, 53, 63, 71–72, 77, 119–23, 139–40, 141, 175–76, 182–94, 199, 211–12, 213, 221, 240n–42n, 246n
neuroplasticity, 21–35, 48–50, 75, 115, 119–20, 139–43, 176, 182–88, 192–97, 211–12, 213, 221–22
neuroscience, 26, 34–35, 37, 51–52, 63–64, 74, 120–22, 126–29, 134–35, 139–43, 175, 182–97, 213, 214–16, 220–21, 240n–41n
neurotransmitters, 19–20, 27, 35, 49, 184–87, 194, 240n–42n
New Republic, 103
news aggregators, 132
news crawls, 131
newspapers, 11, 14, 16, 44, 70, 84, 87, 91, 93, 95, 101, 103, 109, 110, 117
Newsweek, 95, 102, 172
New York, 221
New York Philharmonic, 96–97
New York Public Library, 97–98, 162
New York Times, 95
New York Times Book Review, 102
next-generation architecture, 159
Nexus One, 92
nGenera, 9
Nicholas Nickleby (Dickens), 103

Nielsen, Jakob, 134–35
Nielsen Co., 87, 227
Nietzsche, Friedrich, 17–19, 31, 45, 76, 209, 265n
1984 (Orwell), 252n
nonfiction, 103–4, 198–200
nonlinear reading, 136–38
nonsense words, 182–84
nonverbal IQ performance, 147
Northwestern University, 108
Norway, 146
notebooks, 166–67, 168, 178–79
novels, 74, 103, 104–5
Novum Organum (Bacon), 69
number systems, 116
numbness, 209–10, 213

objective knowledge, 128
object rotation, 141, 145, 147, 148
obsessive-compulsive disorder, 35
octavo, 70
"Ode to Psyche" (Keats), ix
O'Faolain, Sean, 127–28
office computers, 132
oil-based ink, 68
Old Order Amish, 47
Olds, James, 26
Old Testament, 210–11
"On Computable Numbers, with an Application to the Entscheidungs-problem" (Turing), 81–82
one-time reading, 138
Ong, Walter J., 51, 56, 57, 77, 246n, 248n
Onishi, Norimitsu, 104–5
Online Book Pages, 163
online services, 14
on-screen editing, 13
On the Genealogy of Morals (Nietzsche), 265n
On the Origin of Species (Darwin), 76
operating systems, 14, 113, 159, 172
oral culture, 53–57, 60, 67, 72, 77, 177–78, 248n
Orality and Literacy (Ong), 57
O'Reilly, Tim, 105
O'Reilly Media, 105
organic model, 179, 181–87, 190–93, 195
"orphan books," 163
Orwell, George, 252n
O'Shea, Joe, 8–9, 138
Ostrosky-Solis, Feggy, 51

outsourcing, 191–97
Oxford University Press, 162

Page, Larry, 152–56, 159, 161–62, 171–72, 175
pain, 220–21
Palo Alto Research Center (PARC), 113–14
Pandora, 200
paper, 59, 69, 88, 126–27, 166
paper documents, 126–27
Papinian, 70
papyrus, 59, 60
paralysis, 30
parchment, 59, 60, 68, 69, 166
parietal cortex, 213
Paris, 70
Paris, University of, 134
parsing, 61–62
Parts of Animals, The (Aristotle), 36–37
Pascual-Leone, Alvaro, 31, 33
pastoral ideal, 166–67
Pastoral Symphony (Beethoven), 96
pathology, 34–35
patriarchy, 126
PC Magazine, 154
Pearson PLC, 223
peer pressure, 118
Penfield, Wilder, 24
peripheral nervous system, 25
peripheral vision, 29
personal cataloguing machines, 169–70
personal computers (PCs), 4, 11–16, 83, 88, 113, 144, 148, 169–70, 172, 260n
personal memory, 196–97
Phaedrus, 54–55
Phaedrus (Plato), 54–57
"phantom limbs," 29–30
Philadelphia Inquirer, 93
Phoenician alphabet, 53
phonetic alphabet, 51, 53–54, 55, 57, 62
phonograms, 109
"phonographoteks," 110
phonographs, 77, 88, 95, 109, 110, 219–20
phonography, 110
photocopiers, 180
photography, 83, 85, 89, 159
photo storage, 159
physical therapy, 30
Piaget, Jean, 40
piano playing, 33

Pilzecker, Alfons, 183–84
pineal gland, 23, 37
Pines, Maya, 64
Pi Sheng, 246n
pixels, 100, 103
"plan-based behavior," 215
plasticity, brain, 21–35, 48–50,
    75, 115, 119–20, 139–43, 176,
    182–88, 192–97, 211–12, 213,
    221–22
Plato, 54–57, 66, 67, 69, 70
plows, 44, 45, 210
pocket watches, 43
podcasts, 85
poetry, 55–56, 73–74, 75, 111
Poldrack, Russell, 133
Popular Mechanics, 80
Popular Passages, 165
population growth, 48
pop-up ads, 117
"portrait" mode, 116
positive reinforcement, 117
Postal Service, U.S., 83, 92
postcards, 92
posterior hippocampus, 32–33
"post-literary mind," 111–12
Postman, Neil, 151–52
postmodernism, 126
postsynaptic neurons, 186–87
power, balance of, 48
power browsing, 136–38
power looms, 212
PowerPoint, 105
"practice" events, 119–20
prayers, 42
Preface to Plato (Havelock), 55
prefrontal cortex, 122, 213, 246n
Prelude, The (Wordsworth), 76
presynaptic neurons, 186–87
prices, book, 67, 69, 100–101
primary memories, 183
primates, 32, 186, 208
Princeton University Press, 162
Principles of Psychology (James), 21
Principles of Scientific Management, The
    (Taylor), 150
Pringle, Heather, 6
printing, 2, 10, 68–72, 76, 77, 83–84,
    89, 94, 97, 99–100, 102, 110–11,
    162, 168, 177–78, 246n
Printing Press as an Agent of Change,
    The (Eisenstein), 69–72

problem solving, 119, 122–26, 146–48,
    214–16
Prodigy, 14
productivity, 41, 149–50
programmers, computer, 11, 12, 100,
    113, 150, 157, 201–8, 216, 224
progressive education, 180
"Project for a Scientific Psychology"
    (Freud), 21–22
prose style, 66
proteins, 184–85, 191
Proust, Marcel, 111
Proust and the Squid (Wolf),
    51–52
PSAT scores, 145
psychological development, 19, 64, 65,
    79–80, 117, 132, 151–52, 166–67,
    168, 171, 173, 204–5, 209–10, 212,
    213, 219–21
Psychological Science, 74, 219
psychotherapy, 19, 204–5
public address systems, 79
public libraries, 97–98
public schools, 44, 92–93, 126–27, 146,
    163, 192–93, 223
publishing industry, 2, 10, 15–16,
    66–72, 76, 77, 83–84, 87, 89, 92,
    94–95, 97, 99–114, 157, 161–66,
    168, 177–78, 246n
punctuation marks, 62
puzzles, 148, 214–16
Pygmalion (Shaw), 202

quarto, 70

Rabelais, François, 70
radio, 77, 78–80, 83, 84, 87, 89, 93, 99,
    110, 112, 117
railroads, 167, 219
RAM, 12
Ramachandran, V. S., 29–30
Ramón y Cajal, Santiago, 22
rationalism, 28–29
rats, 263n
reading, 5–8, 50–77, 87–88, 90, 97,
    106–12, 120, 122–29, 134–38, 140,
    141, 168, 177–78, 199, 201, 246n,
    253n
    see also books
realism, 40
real-time updates, 158
reasoning, 43–44, 50, 140, 147

reductionist fallacy, 175–76, 181–82
reference books, 66, 88, 92–93
reflex response, 139
regeneration, 49–50
relative knowledge, 128
Relativity (Einstein), 76
"relevant content," 166
religion, 42, 47, 50, 60–68, 152, 174, 181, 251*n*
Renaissance, 10, 43, 165, 179, 181
repetition, 116, 184–85
Republic, The (Plato), 56
Republic of Letters, 72, 247*n*
retrograde amnesia, 183
Revolution in Time (Landes), 42
Reynolds, Joshua, 142
Rich, Barnaby, 168
Rich, Motoko, 101
right-handedness, 31–32
Rin, 105
ringtones, 91, 94, 117
rituals, 218–19
Robinson, Arthur, 41
Rocky Mountain News, 93
Rogerian psychotherapy, 202–5
Rogers, Carl, 203
Rolling Stone, 94
Roman alphabet, 53
Roosevelt, Franklin D., 169
Rosen, Christine, 103
Rosenblum, Kobi, 191
roughtype.com, 15
RSS, 16, 91, 157–58, 159, 199, 200
Rummel, Erika, 179

saccades, 134
Saenger, Paul, 61, 63, 66
Sagan, Carl, 205
salience network, 232–33
Salomé, Lou, 265*n*
Samsung Group, 96
samurai culture, 47
Sarnoff, David, 3–4, 46
satellite television, 96
SAT scores, 146
Scandinavia, 146
scanning, visual, 134–37, 141–42
scans, brain, 26, 37, 51–52, 74, 175, 213, 220–21
scans, page, 161–66
schemas, 124–25, 216
Scherer, Michael, 95

Schmidt, Eric, 97, 150, 152, 162, 173
Schoeffer, Peter, 69
scholarship, 111–12, 153–54, 217–18
schools, 44, 92–93, 126–27, 146, 163, 192–93, 223
Schulman, Ari, 176, 195
Schwartz, James, 186–87
Schwartz, Jeffrey, 35
Schwarzenegger, Arnold, 93
Science, 141, 147, 217
scientific progress, 23, 40–41, 45, 70, 76, 147, 149–50, 169–70, 211, 217–18
"scientific spectacles," 147
screens, computer, 12, 14, 77, 87, 94, 100, 103, 107, 116, 138, 169
scribes, 60, 61, 62, 65, 67, 69, 89, 94
Scribner's Magazine, 109–10
scripts, automated, 202–5, 213, 218–19, 223–24
scriptura continua, 61, 65, 122
scrolling, 116
Scrolling Forward (Levy), 73, 113–14
scrolls, 59, 67, 164
search boxes, 117–18
search engines, 6, 90–91, 106, 132–33, 181, 216
  see also Google
sea slug, *see Aplysia*
Seattle Post-Intelligencer, 93
secondary languages, 176
secondary memories, 183
"secondary referencing," 137
seizures, 183, 188–89
self-employment, 199
Seneca, 141, 179
sensory-motor experiences, 90, 188–89
sensory nerves, 24–26, 29–30, 44, 64–65, 122
serotonin, 186, 187, 241*n*–42*n*
sextants, 44
Shakespeare, William, 70, 178
Shallows, The (Carr), writing of, 198–200, 223
Shaw, George Bernard, 202
Shirky, Clay, 111–12
shopping, 9, 89, 130
shortcuts, 95
short-term memory, 123–26, 182–97
Siegel, Daniel, 190
"signals," 158
silent reading, 60–62, 63, 66, 67, 73–74, 97, 106–7

Silent Spring (Carson), 76
silicon chips, 23, 83, 180
silicon memory systems, 180
Silicon Valley, 83, 150, 153, 154
Simon & Schuster, 105–6
Singhal, Amit, 158
"skimming activity," 134–37
Skype, 85, 199
slang, 252n
sleep, 190, 262n–63n
Sleepy Hollow, 166–67, 168, 220
slide rules, 44
Small, Gary, 120–22, 125–26, 139–40
smartphones, 15, 92, 97, 159, 226–27, 228–38
    cognitive and emotional effects of,
    228–29
"snippets," 161, 165–66
Snow White, 96
social network, 6, 85–86, 91, 96, 97, 98,
    106–7, 117–18, 132–34, 158, 159,
    198, 200, 248n, 251n
societal complexity, 147
Socrates, 54–57, 67, 177, 195
software, computer, 13–14, 82, 83, 84,
    91, 100, 107, 113, 132, 135, 150,
    157–58, 159, 170, 175, 195, 201–8,
    214–19
Software Usability Research Laboratory,
    135
solid-state electronics, 79
somas, 19–20
somatosensory cortex, 26, 116–17
Sony Corp., 96
sound-processing equipment, 84
soundtracks, 85
South Korea, 196
Space between Words (Saenger), 61
Spain, 70
Sparrow, Betsy, 235
spatial representation, 32–33, 39–41,
    49–50, 53, 211–12, 262n–63n
Speer, Nicole, 74
split A/B testing, 151
Spotify, 94
spreadsheets, 13, 159
standardization, 42–43
Stanford University, 141–42, 153–54, 172
Star Wars, 11
static pages, 157
steam engine, 149
steam mills, 46
Steiner, George, 111–12

Stevens, Wallace, 73–74
Stone, Brad, 101
Stone Age, 49
storage devices, 159, 180–82, 191–92
Strategy & Business, 97–98
strokes, 30
Structure of Scientific Revolutions, The
    (Kuhn), 76
stylus, 59–60
subjectivity, 151, 213, 220–24
subject lines, 117
subsistence economies, 147
Suderman, Peter, 181
Sumerian language, 52–53, 58–59
sundials, 41
"supernormal stimulus," 233
surfing, Internet, 87, 96, 115–16,
    121–22, 130–31, 150–76
"survival of the busiest," 35, 142
Sweller, John, 123, 131
"switching costs," 133
symbols, 55, 83–84, 118, 148, 175
synapses, 19–22, 26–27, 34–35, 49,
    71–72, 120, 141, 182–93, 194, 199,
    240n–42n
synaptic consolidation, 188
Synaptic Self (LeDoux), 28–29
synchronization, 42
syntax, 56, 61, 62, 63, 75, 107, 201, 253n
synthesis, 178–79
"sysprogs," 12
system consolidation, 188

tables of contents, 90
tablets, 52–53, 58–60, 67
tabula rasa, 28
tape decks, 84, 88
Tapscott, Don, 144–45, 181
Taub, Edward, 30, 31–32
taxi drivers, 32–33, 49–50, 188, 212
Taylor, Frederick Winslow, 149–50, 152,
    218
Taylorism, 149–50, 152, 173, 218
Technics and Civilization (Mumford), 44
technological determinism, 46
technology:
    binary, 94, 190–91, 199
    categories of, 44–45
    communications, 78, 79, 83, 84, 88,
    133–34
    cultural influence of, 2–16, 38,

44–50, 68–72, 76–77, 85–98,
  101–7, 138, 194–97, 201–19
data-processing, 16, 82–83, 133–34,
  153–54, 167–70, 180, 192–93, 208
determinist vs. instrumentalist
  approach to, 46–48
development of, 46–48, 88–90
digital, 8–9, 79, 81–82, 83, 88–89,
  94, 100, 101, 160, 161–66, 190–91,
  199, 207, 214, 248n, 252n–53n
entrenchment of, 221–22
information, 1–4, 44–45, 81–82,
  116–20, 123, 138–43, 146
intellectual, 23, 44–46, 50–51,
  82–83, 138–39, 150–52, 161–66,
  180–81, 191–219, 221–22
investment in, 154, 155
mobile, 77, 86, 87, 91, 96–97, 101,
  102–5, 116, 117
personalization of, 4, 11–16, 43, 83, 88,
  113, 144, 148, 169–70, 172, 260n
uncritical belief in, 44, 99–114, 152,
  175–76, 181–82, 210–12, 260n
see also specific technologies
Technopoly (Postman), 151–52
telecommunications, 78, 79, 83, 84, 88,
  133–34
telegraph, 78
telephones, 11, 79, 83, 84, 88
television, 11, 77, 83, 84, 85, 86–87, 88,
  93, 94, 95–96, 110, 112, 131, 144,
  200, 260n
templates, mental, 28
temporal filtering, 170–71
temporal lobe, 51–52, 213
temporal representation, 41–44, 45, 50,
  70, 211
Terence, 178
terminals, computer, 11–12, 201
testosterone, 241n–42n
textbooks, 92–93, 131
text displays, 77, 90, 91–92, 105–6,
  128–38, 139, 161–66, 217–18
text medium, 1, 2, 5–10, 14, 53–57,
  94–95, 99–114, 246n
  see also books
text messages, 86, 96–97, 104–5, 108,
  117, 120, 132, 252n–53n
Thamus, 54
theme channels, 96
Theuth, 54
Thompson, Clive, 6, 180, 182

Thompson, Hunter S., 94
Thomson, David, 4
three-dimensional games, 85
Tigris River, 52
Time, 251n
time-and-motion studies, 149–50
Times (London), 95
Times Education Supplement, 223
time-sharing, 11–12, 204–5, 207
tinnitus, 35
tokens, 52
Tolstoy, Leo, 7, 111, 112
topography, 39–41
touch-sensitive screens, 116
touch typing, 17–19
trackpads, 116
tractors, 210
Tracy, James, 260n
Transcendentalism, 166–67
transcranial magnetic stimulation
  (TMS), 33
transistors, 79
translucent screens, 169
"transparent eyeball," 166
trauma, brain, 30–31, 34, 183, 188–89
triode tubes, 78–80
"Trout, The" (O'Faolain), 127–28
Turing, Alan, 81–82, 174
Turing machine, 81–82, 102, 202, 205–6
Turing test, 205–6
Twitter, 85, 91, 96, 97, 117, 132, 158,
  159, 198, 251n
two-pole vacuum tubes, 78–79
2001: A Space Odyssey, 5, 24, 172–73,
  175, 224
typewriters, 17–19, 44, 45, 81, 83, 109,
  209, 265n
typography, 68, 83–84, 101, 102, 117

"unbundling," 94
unconscious mind, 119
*Understanding Media: The Extensions of
  Man* (McLuhan), 1–2, 46, 89, 210
"universal machines," 81–82, 102, 202
Universal Studios, 96
universities, 81, 92, 136–37, 147, 162,
  168–69
University College London, 136–37
updates, news, 158
Updike, John, 104
"upgrade cycle," 14
urbanization, 42–43, 147, 219–20

usability lab, 151
Uzanne, Octave, 109–10

vacuum tubes, 78–80
van Nimwegen, Christof, 214–16
Veblen, Thorstein, 46
venture capital, 154, 155
verbal scores, 145–46, 223
vernacular languages, 62
Vershbow, Ben, 106
video, 84–85, 91–92, 105–6, 116–17,
    129–34, 180, 200
video games, 12, 13, 85, 98, 139, 144
videotapes, 180
vinyl records, 89
violinists, 31–32
Virga, Vincent, 40
vision, 29, 189
visual cortex, 29, 63, 116–17, 189
visual processing, 29, 51, 63, 116–17, 121–
    22, 129, 131, 134–36, 139–43, 189
visual-spatial intelligence, 141, 144–48
"vital paths," 22, 27
Vizplex, 100
vocabulary, 75, 107, 145, 146, 253n
voice calls, 86
von Neumann, John, 174, 176
"vooks" (video books), 105–6
vowel sounds, 53

Wallace, David Foster, 194–95
Wall Street Journal, 152, 162, 221
War and Peace (Tolstoy), 7, 111, 112
Ward, Adrian, 229, 234–237
Washington Post, 93
Watchmen, 96
wax tablets, 59–60
weaponry, 47, 48
weaving, 210
webcams, 85
Web design, 115–16, 159
Web hosting, 159
Web pages, 83–84, 91, 115–16, 120,
    135–36, 157, 159, 218–19
Web publishing, 157
Web sites, 9, 83–84, 91, 93, 94, 120,
    135–36, 153, 154, 157–58, 159, 165,
    216–19
Wegner, Daniel, 235, 237
Weisberg, Jacob, 102
Weizenbaum, Joseph, 201–8, 211,
    223–24

Wells, H. G., 49
Wells, Jonathan, 93
Wenner, Jann, 94
Westwinds Community Church, 251n
What Is Intelligence? (Flynn), 146–47
White, Lynn, 43
Whitman, Walt, 196
"Why Minds Are Not Like Computers"
    (Schulman), 176
Widener Library, 162
widgets, 91
Wi-Fi, 15, 97, 200
Wikipedia, 15, 85, 102, 103
windmills, 46
Windows operating system, 14, 113, 172
Winner, Langdon, 47
Wired, 14
wireless transmission, 78–80
Wolf, Maryanne, 51–52, 53, 63, 75–76,
    122
Wolf Trap Center, 96
women, 104–5, 139
woodblock printing, 89, 246n
wooden-screw presses, 68
word-processing programs, 12, 13, 16,
    152, 159, 209–10
Wordsworth, William, 22, 76
working memory, 123–26, 131, 132–33,
    139, 141–42, 182–97
World Almanac, 88
World Brain (Wells), 49
World War II, 81
World Wide Web, see Internet
wristwatches, 43
writing, 50–77, 103, 104–8, 145–46,
    170–71, 178–79, 198–200, 209–10,
    223, 246n, 247n, 252n–53n
writing balls, 17–19, 45, 265n
writing skills, 145–46

Xerox Corp., 113–14
X Generation, 10–11

Yahoo.com, 15, 16
Yellow Pages, 89
Young, J. Z., 21, 72
YouTube, 15, 85, 94, 96, 160, 200

Zhu, Erping, 128–29
Zip drives, 14
Zuckerberg, Mark, 158